OCEANOGRAPHY AND OCEAN ENGINEERING

BALTIC AND BLACK SEA

ECOLOGICAL PERSPECTIVES, BIODIVERSITY AND MANAGEMENT

OCEANOGRAPHY AND OCEAN ENGINEERING

Additional books in this series can be found on Nova's website under the Series tab.

Additional e-books in this series can be found on Nova's website under the eBooks tab.

OCEANOGRAPHY AND OCEAN ENGINEERING

BALTIC AND BLACK SEA

ECOLOGICAL PERSPECTIVES, BIODIVERSITY AND MANAGEMENT

ANDRE MILLS
EDITOR

nova publishers
New York

NOTICE TO THE READER

Library of Congress Cataloging-in-Publication Data

Names: Mills, Andre, editor.
Title: Baltic and Black Sea : ecological perspectives, biodiversity and management / editor, Andre Mills.
Description: Hauppauge, N.Y. : Nova Science Publisher's, Inc., 2016. | Series: Oceanography and ocean engineering | Includes index.
Identifiers: LCCN 2016017091 (print) | LCCN 2016021991 (ebook) | ISBN 9781634852333 (hardcover) | ISBN 9781634852456 (Ebook) | ISBN 9781634852456 ()
Subjects: LCSH: Baltic Sea--Environmental conditions. | Black Sea--Environmental conditions. | Marine biodiversity--Baltic Sea. | Marine biodiversity--Black Sea.
Classification: LCC GE160.B29 B34 2016 (print) | LCC GE160.B29 (ebook) | DDC 577.7/334--dc23
LC record available at https://lccn.loc.gov/2016017091

Published by Nova Science Publishers, Inc. † New York

CONTENTS

PREFACE

This book discusses ecological perspectives, biodiversity and management of the Baltic and Black Seas. Chapter One reviews the role of the Russian section of the Gulf of Finland and its basin in the ecosystems of the Baltic Sea. Chapter Two studies spatio-temporal dynamics of ichthyoplankton biodiversity in different biotopes of the Baltic Sea. Chapter Three explores the abnormalities in the amphids of free-living nematodes in the Black Sea. Chapter Four discusses the role of defense systems to the adaptation to ecological conditions in the Black Sea ecosystem.

Chapter 1 - Russia occupies a small part of the Baltic Sea, but it still exerts influence upon biodiversity of the whole Sea and out of its limits: 1. a part of spawning grounds of fishes and lampreys migrating over most of Baltic Sea are located in the Russia. Meanwhile some populations are not supported by hatcheries, natural reproduction of them still takes place. 2. In the Russian section of the Gulf of Finland and its basin several refuges for the wildlife exist. The populations of threatened animals survived there (freshwater pearl mussel, European mink). 3. During spring migrating birds move along the eastern coast of the Sea to the north up to "dead end" – narrow end of the Gulf of Finland. Big congestions of birds concentrate there. In the wetlands of this area birds stop for some time, relax, feed, and then rapidly migrate to the Russian Arctic and other inland territories for the nesting. The back-migration of birds in autumn is less concentrated in respect to time and space. The condition of eastern extreme of the Baltic Sea determines the condition of bird populations of a big territory. Nowadays the wetlands of this extreme are threatened, because most of them are located close to urban area. 4. The negative aspect of the influence of the eastern part of the Baltic Sea consists in the fact, that it is the source of invasions of alien species.

Chapter 2 - The Baltic Sea could be considered as an example of the successful ecosystem functioning under the small biodiversity of biota. The long-term data on the ichthyoplankton species diversity of the south-eastern Baltic gives an opportunity to examine its dynamics on spatio-temporal aspects. The deep-water part and shallow zone of the Baltic Sea are sharply distinguished by species composition, abundance and ecological adaptations of egg and larvae assemblages. Only 4 fish species with pelagic eggs (cod, sprat, flounder, rockling) spawn in the deep basins of the Eastern Baltic. The scarcity of species richness could be compensated by specific adaptations: the extremely prolonged spawning season (cod, sprat, rockling), a transition to different types of spawning biotopes in the course of reproductive season (sprat) or in different areas of the Baltic (flounder). Under the invariable species richness through several decades, the Shannon's index of general diversity demonstrated a tendency to decline with minimum in the late 1990s - early 2000s due to dramatic environmental changes. Currently, early developmental stages of sprat occupy the dominant position in the deep-water ichthyoplankton assemblage of the eastern Baltic. The spawning habitats in the coastal shallow zone are used mainly by fishes laying the bottom eggs. Larval assemblages included 22 species. This increase of species richness was defined by the bottom substrate diversity, the seasonality of temperature regime, depth gradients and alternations of upwelling and downwelling processes. Long - term changeability of species richness could be traced between 1946-1955 and 1992-2007 based on literature sources and field investigations. In the coastal zone a trend was also observed to decline of general diversity indices. The one of tendencies was a diminution of occurrence and habitat fragmentation of larval species from winter spawning (snakefish, rock eel) on contrary with dominance of fish larvae from summer reproduction (gobies). A mutual trend was revealed to transition from polydominant to monodominant assemblages both in deepwater basins and shallow coastal zone. On the open part of the Baltic this was linked to abundance of early developing stages of small pelagic sprat while in the coastal zone with dominance of small short-lived bottom species, a namely, sand goby. These tendencies were linked to the climatic and environmental changes during last decades.

Chapter 3 - In this chapter the authors aim to summarize their long-term data (1988-2015) about the abnormalities detected in the development of the vital chemoreceptor organs (amphids) of 49 nematode species from the Black Sea. The total list of species includes 28 valid species and 21 morphotypes, systematic position of which have been identified only to genus level as well as individuals close to a valid species (cf.), however, they require further study and clarification in the authors' view. Amphid abnormalities are characterized by a wide spectrum of appearances. Accordingly, they may have a non-standard number of elements of a species-specific form (instead of two elements they may have 1 or 3-5); one element of amphid may be underdeveloped or not have a species-specific form; the topography of amphids in relation to the front edge of the head and to the axis symmetry of the body may be disturbed. Anthropogenic impact on the Black Sea ecosystem is steadily increasing. Its negative effect is revealed in the extinction of species, community perturbations, serious physiological changes in organisms, resulting in different pathological phenomena. It is obvious that the occurrence of specific abnormalities in the vital organs of the same nematodes species is a result of chromosomal mutagenesis. Repeatability and the wide distribution of species with these types of abnormalities in different regions of the basin are shown in this chapter.

The issues indicated in this chapter require future research: 1. what is the specific cause of the amphid abnormalities in nematodes? 2. Are these amphid abnormalities offer an advantage or a disadvantage compared to the "normal" organs for organisms inhabiting polluted ecosystems or a life under the impact of climate change? 3. Why similar morphological abnormalities are not recorded in nematodes from the other seas? Are these chemoreceptor organ abnormalities in nematodes unique for the Black Sea? 4. Is this phenomenon shows the nascent process of microevolution of nematode fauna in the Black Sea?

Chapter 4 - Black Sea is one of the largest indoor anoxic sea of the world. The specific ecological conditions of the Black Sea, including anoxic water containing high concentrations of the hydrogen sulfide, play an important role in fish adaptations to unfavorable natural factors and need to study the adaptive mechanisms of their inhabiting in different biotops. In addition, great pollution level in the Black Sea caused high anthropogenic activity, especially in the coastal area, influences directly and indirectly on fish populations and their health. Assessment of fish natural populations health is a great problem because normal values range widely and depend to natural ecological conditions and their fluctuations. Biomarker analysis of field-collected species can provide information on their physiological status and environment. Therefore, the main idea of the chapter is to study the role of blood biomarkers of highly distributed Black Sea fish species belonging to different ecological groups to the different natural and anthropogenic impacts on marine environment, to determine their main trends and importance for fish adaptations for keeping their life with the changes of ecological conditions. For this purpose the biochemical parameters of various highly distributed Black Sea fish species inhabiting different biotops were studied and the responses of biomarkers were detected. The levels of heavy metals in examined fish species tissues were determined and the responses of blood biomarkers were observed. The biomarkers studies help in understanding the relationship of blood biochemical characteristics to the habitat and adaptability of the fish species to the environment. The principles and perspectives of the development of permanent monitoring programs in the Black Sea for fish biodiversity protection used biomarkers are discussed.

In: Baltic and Black Sea
Editor: Andre Mills

ISBN: 978-1-63485-233-3

Chapter 1

ROLE OF THE RUSSIAN SECTION OF THE GULF OF FINLAND AND ITS BASIN IN THE ECOSYSTEMS OF THE BALTIC SEA

Igor Popov*
Saint Petersburg State University,
St. Petersburg State University, St. Petersburg, Russia

ABSTRACT

Russia occupies a small part of the Baltic Sea, but it still exerts influence upon biodiversity of the whole Sea and out of its limits: 1. a part of spawning grounds of fishes and lampreys migrating over most of Baltic Sea are located in the Russia. Meanwhile some populations are not supported by hatcheries, natural reproduction of them still takes place. 2. In the Russian section of the Gulf of Finland and its basin several refuges for the wildlife exist. The populations of threatened animals survived there (freshwater pearl mussel, European mink). 3. During spring migrating birds move along the eastern coast of the Sea to the north up to "dead end" – narrow end of the Gulf of Finland. Big congestions of birds concentrate there. In the wetlands of this area birds stop for some time, relax, feed, and then rapidly migrate to the Russian Arctic and other inland territories for the nesting. The back-migration of birds in autumn is less concentrated in respect to time and space. The condition of eastern extreme of the Baltic Sea determines the condition of bird populations of

* Email: i.y.popov@spbu.ru.

a big territory. Nowadays the wetlands of this extreme are threatened, because most of them are located close to urban area. 4. The negative aspect of the influence of the eastern part of the Baltic Sea consists in the fact, that it is the source of invasions of alien species.

Keywords: anadromous fishes, threatened animal species, alien species, migrating birds

INTRODUCTION

Russia occupies the eastern part of the Gulf of Finland of Baltic Sea. This part reminds a big lake, because the water is almost fresh there. A big city Saint-Petersburg is situated at its eastern extreme. The population of the city exceeds five million people. This results in the high anthropogenic pressure upon surrounding territory. However the Russian section of the Gulf of Finland and its basin still exerts influence upon biodiversity of the whole Sea and out of its limits. The results of studies of birds, mammals, fishes and molluscs provide especially remarkable evidences on such an influence. Recently these studies have been conducted in a framework of projects on protected areas (Aksenov et al., 2013) and fishery management (Popov, 2010, 2014). This article presents a review of the most interesting results in this field. The own data focused mainly on aquatic species and some mammals. The other data have been collected from the publications in local editions.

SPAWNING GROUNDS OF THE ANADROMOUS FISHES AND LAMPREYS

Baltic Sea is relatively poor in respect to fish fauna. The main part of fish biomass is composed by small number of fish species. Some of them are anadromous: they reproduce in rivers, and grow mainly in the Sea. Meanwhile some fishes migrate over whole Baltic Sea. These species are threatened in Europe, because almost all rivers are blocked by numerous dams hampering fish migrations and suffer from other negative anthropogenic impacts. Usually the anadromous fishes spawn in the sections of rivers with strong current and pure ground of bottom. The suitable habitats are often threatened by dredging and pollution. This process takes place in Russia and other countries surrounding Baltic Sea, but in the Russian section the suitable habitats still occupy big areas.

One of the most significant fish species in the Baltic Sea is an Atlantic salmon, *Salmo salar*. It reproduces naturally in two rivers flowing into the Russian section of the Gulf of Finland. The main part of the spawning grounds is located in the Luga River. The riffles suitable for the spawning of the Atlantic salmon occupy several ten thousand square meters there. They are located in the mainstream of Luga (which is 362 km long) and four inflows. Many riffles are surrounded by forests and look like pristine environment (Figure 1). However because of long fishing and overfishing the number of local salmons is very small. During recent studies the salmon parrs of natural origin have been noted only in one riffle of one Luga inflow (Vruda) (Popov, 2001; Popov, 2010).

Figure 1. Spawning grounds of the Atlantic salmon in Luga River.[1]

Spawning grounds of the Atlantic salmon also exist in the system of Chiornaya River flowing at the northern perimeter of Saint-Petersburg city. The salmons have been exterminated there some decades ago, but since 2000 the restocking takes place there. Several thousands of reared salmon parrs have

[1] Most of the photographs in the text have been made by the author over last 5 years, two of them have been presented by colleagues (indicated below).

been released there each year. At least one fact of natural reproduction of salmon was noted there in 2006 (salmon redd with fries), one spawner was caught in 2003 (Popov, 2015).

In the past Atlantic salmon reproduced in two other rivers as well – Narova (Narva) and Neva. The first one is blocked by dam close to the mouth, in the other the spawning grounds have been destroyed by the dredging. In 1970s the shallow riffle in Neva has been deepened to improve the conditions the navigation (Nezhikhovsky, 1981). Since that time no cases of natural reproduction of salmon are known (Popov, 2010). Nowadays both populations exist due to the hatcheries. Neva population is of special value, because it is the only population of the Baltic Sea, which still keeps its genetic uniqueness. All other populations are of mixed origin, because in the past the releases of reared salmons often took place without taking into account the genetic individuality. In Neva River the alien salmons had never been released. Since Atlantic salmon has a strong homing instinct, the migrations of non-native salmon into the rivers take place very rarely. (As for Neva River, during 7 years of observation only one case of the occurrence of Narova salmon is known). The population of Luga river is similar to Neva one in this respect, because just a small number of salmons from neighbour rivers have been released there.

Now the local hatcheries produce annually several hundreds of thousands of the Atlantic salmon juveniles. The tagging showed that they migrate almost over whole Baltic Sea. The main part of them is caught at the coasts of Finland (Kazakov, 1985). The migrations of the other anadromous fishes from the eastern part of the Gulf of Finland are not well studied, but at least some of them – brown trout, *Salmo trutta*, lamprey, *Lampetra fluviatilis*, white fish, *Coregonus lavaretus*, – also can migrate for long distances. Spawning grounds of these species also occupy big areas in the rivers flowing into the Russian section of the Baltic Sea. There are at least 15 rivers and brooks, in which the captures of brown trout had been registered (Popov, Ostrovsky, 2014) (Figure 2). Usually in the same rivers, where either salmon or trout reproduces, the spawning of lampreys also takes place. Spawning grounds of white fishes are not well studied, but most likely they exist at the mouths of the biggest rivers (Neva, Narova, Luga), because these fishes are often caught there (Popov, unpublished data) (Figure 10). In the past the Atlantic sturgeons, *Acipenser sturio*, had been captured sometimes in Luga and Neva (Grimm, 1889; Berg, 1935; Pravdin, 1948). The sturgeons had been exterminated completely in the whole Baltic Sea. However since the rivers Neva and Luga are not blocked by dams, the restocking of sturgeons is possible there.

Figure 2. Spawning grounds of brown trout and lamprey (Ushkovsky brook at the northern perimeter of Saint-Petersburg city).

The conservation of threatened riverine habitats is related with pattern of river management in Russia. Rivers and their banks cannot be private property. The use of them is restricted by series of bans. According the "Water code of Russia" (03. 02. 2006. № 74-FZ www.pravo.gov.ru) "water protective zones" and "coastal defensive stripes" exist at the banks of all water bodies. These notions point out different patterns of bans at different distance from the border of water body. "Coastal defensive stripe" is usually 30-50 m, the "water protective zone" is 30-200 m. In the water protective zone the following activities are not allowed: 1) use of drainage water for fertilizing of soil; 2) allocation of cemeteries, burial grounds for animal refuse, waste burials; 3) aerial pest control; 4) movement and parking of transport outside the roads and specially arranged places. In the water defensive stripes ploughing up, allocation of ground and pasture are not allowed. This results in the fact that at least small stripe of natural arboreal vegetation usually exists along the rivers. This prevents erosion of soil and washout of ground into rivers. In the past the dams have been often built in small rivers to make mills and electric station. However during the existence of Soviet Union all electric system of the country was united and centralized, the small electric stations became unnecessary. The main part of electricity is produced by big stations. In such a situation many dams in small rivers became abandoned and destroyed, that is why many rivers still have connections with a Sea.

REFUGES FOR THE WILDLIFE

Wildlife areas and the populations of threatened species are still relatively numerous in the Russian section of the Gulf of Finland and its basin (Figure 10). A recent windfall of the freshwater pearl mussels, *Margaritifera margaritifera*, was especially surprising fact (Ostrovsky, Popov, 2008). This bivalve species suffers a dramatic decline in Europe. About 90% of its populations became extinct over last decades. Many surviving populations do not reproduce naturally for a long time (Araujo, Ramos, 2001), the artificial rearing was organized in the most of European countries to prevent the extinction of this species (Tielen, 2011). This decline takes place mainly because of destruction of suitable habitats (pollution, construction of dams, dredging, extermination of natural arboreal vegetation at the river bank causing washout of soil, etc.). During recent studies in five rivers flowing into the Gulf of Finland pearl mussels have been discovered or rediscovered (Popov, 2015). Meanwhile 4 of them flow close to Saint-Petersburg city. In Europe the mussels often disappear in much less populated area. Total number of pearl mussels living close to the Gulf of Finland is about 30000-50000 individuals.

The other endangered bivalve species – thick-shelled pearl mussel, *Unio crassus*, – suffer a decline in Europe like pearl mussel. Both species require specific habitats in small clean rivers. *Unio crassus* also occurs close to Saint-Petersburg. It was found just in three rivers (Popov, Ostrovsky, 2014), but any special research had been carried out, most likely they are more numerous. The other local freshwater bivalves (*Unio tumidus, Unio pictorum, Anodonta cygnea, Anodonta anatina*) are not considered threatened in a global scale. But they have such a perspective. In the past the shells of them were used as row material for the production of buttons, the meat was used as a feed for domestic animals (Znadin, 1938). Now such a use is out of question because of the low number of them. In the basin of the Gulf of Finland in Russia the bivalves are still numerous in some water bodies (Figure 4).

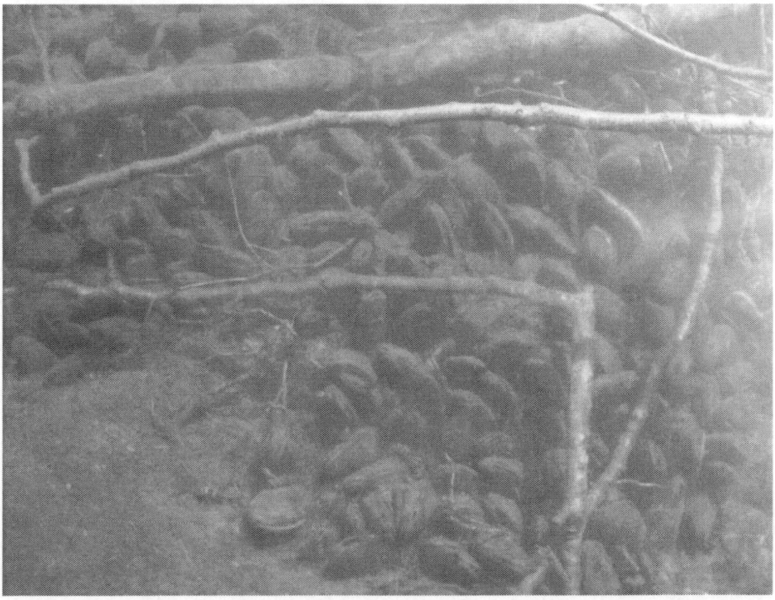

Figure 3. Pearl mussels in a brook (Peipia) at the southern coast of the Gulf of Finland.

Figure 4. Rests of bivalves (*Unio pictorum, Anodonta cygnea*) eaten by muskrats (*Ondatra zibetica*) (Lembolovskoe Lake).

During recent studies noble or broad-clawed crayfishes, *Astacus astacus*, were found close to the Gulf of Finland in Russia. This species is considered now as vulnerable by IUCN (Erdsman et al., 2010). It suffers from non-native disease (crayfish plague), overfishing and competition with alien crayfishes. It is declining in Russia like in European countries, but at least some refuges still exist in some lakes.

Figure 5. Broad-clawed crayfish caught at the Russian-Finish borderline.

European minks, *Mustela lutreola*, have been observed about 10 years ago in the Gogland Island located in the Russian section of the Baltic Sea (Gaginskaya, 2006). Probably they still occur there, because its habitat had not changed significantly. This species became extinct in the most of former habitats because of competition with alien species American mink, *Neovison vison*. In Russia most of European minks survived at the north-eastern extreme of their distribution area (north-eastern part of European Russia) (Skumatov, 2006), but the American mink gradually replaces it there. In such a situation the islands of the Baltic Sea are more effective refuge for the endangered species.

Figure 6. Flying squirrel[2] and its habitat at the northern perimeter of Saint-Petersburg city.

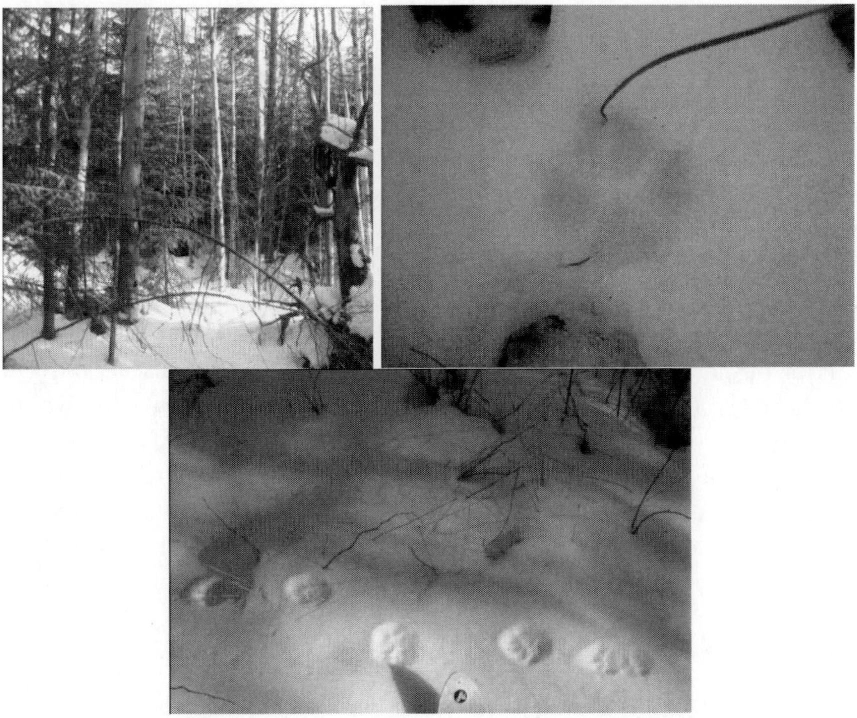

Figure 7. Traces of lynx and its habitat at the southern coast of the Gulf of Finland.

[2] Photograph presented by Alexander Marunchak.

Some refuges for endangered species in the Russian section of the Gulf of Finland basin had been formed because of specifics of forestry. The forests cannot be private property in Russia, that is why their exploitation is rather extensive than intensive. In Europe the forests tend to transform into tree plantation: the amount of dead wood is insignificant there, the old trees or unnecessary for forestry trees are scarce. Many forests have been replaced for monoculture of spruce or other tree species. This is untypical for Northwest Russia, usually the forests renovate by themselves, and they always contain dead wood and other objects, which are unnecessary for forestry. Such a "disorder" created habitats for some animals, which become very rare in Europe – bears, *Ursus arctos*, wolves *Canis lupus*, lynx, *Lynx lynx*, woodpeckers, *Dendrocoppus*, flying squirrels, *Pteromys volans*. Usually, it is possible to find these animals in the forests around Russian section of the Gulf of Finland (Figures 6, 7).

During recent studies interesting data on bats have been obtained at the Gulf of Finland. It turned out that pond bat, *Myotis dasycneme*, which is one of the most rare bat species of Europe, is numerous there (Popov et al., 2009). Probably, there is the biggest pond bat population. 1000-1500 pond bats were calculated annually in hibernacula located at the distance of 100 km eastwards from the Gulf of Finland (Kovalev, Popov, 2011; Kovalev et al., 2015). In summer they were observed mainly in the south-western part of the Russian section of the Gulf of Finland basin. However they spread north-westwards as well. We believe, the big number of pond bats is related with the existence of specific habitats. They usually occupy the small euthrophic lakes (about 50 metres in diameter) surrounded by dense arboreal vegetation or similar small gulfs of big lakes (Figure 8). Especially big congestions of insects form in such habitats. The main threat upon them is an activity to put an order in nature: cutting of unnecessary trees, removing of dead wood, extermination of shrub, intensive grass cutting, purifying of small water bodies from wood and rests of plants, etc. All this reduce the number of insects, as well as the shelters for bats. Such activity continuously progresses as in Europe and Russia, but at least now some "abandoned" ponds or lakes still exist in the territory under study providing habitats for the pond bat. The other bat species also like such places. In the past up to 6 species had been observed simultaneously in such conditions (Kuziakin, 1950). We had not found such cases, but at least three species occurred sometimes (*Myotis daubentoni, Myotis dasycneme, Eptesicus nilssoni*).

Figure 8. Pond bats (cave at the Volkhov River) and their summer habitat at the coast of the Gulf of Finland.

Migration of Birds Through the Russian Section of the Gulf of Finland

During spring numerous birds migrate along the eastern coast of the Baltic Sea to the north-east up to turning out to be in "dead end" - the narrow eastern extreme of the Gulf of Finland. Some of them stop there for some time for feeding and relaxing. Afterwards they rapidly move to the Russian Arctic and other inland areas for nesting. In such a situation congestions of birds concentrate in small areas of the Russian section of the Gulf of Finland. The wetlands of its eastern part and some shallow lakes are especially significant for them (Figure 8). Most of them are threatened, because they are located close to urban area. Several wetlands had been exterminated by dredging and development. However tens thousands of birds still concentrate in the eastern part of the Gulf of Finland during spring (Noskov, 2002). The back migration in autumn is less concentrated in respect to route and time.

Figure 9. Congestion of migrating birds[3] and wetlands at the Kotlin Island, Neva bay.

[3] Photograph presented by Dmitry Starikov.

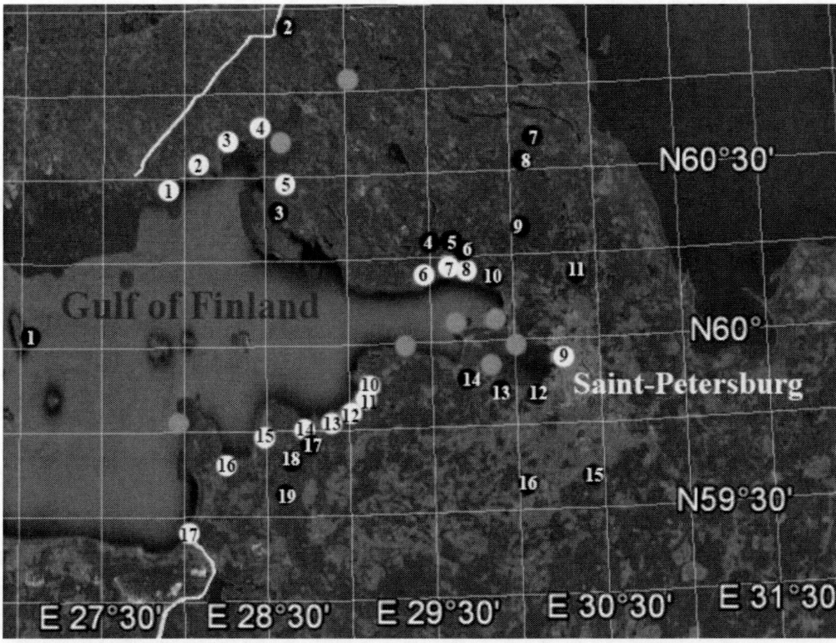

Figure 10. Objects of high conservation value in the Russian section of the Gulf of Finland and its basin: ○ the rivers with spawning grounds of anadromous fishes and lampreys. 1. Serga, 2. Peschanaya. 3. Velikaya. 4. Chulkovka. 5. Gorokhovka. 6. Privetnenka. 7. Chiornaya. 8. Ushkovsky. 9. Neva. 10. Kovash. 11. Voronka. 12. Sista. 13. Koporka. 14. Peipia. 15. Khabolovka. 16. Luga. 17. Narova. ● populations of the threatened species (the species indicated in IUCN red list as vulnerable, endangered or near-threatened) – 1. European mink, 2. Noble crayfish. 3. Pond bat. 4. Pearl mussel. 5. Pearl mussel. 6. Thick-shelled mussel. 7. Noble crayfish. 8. Pearl mussel. 9. Pearl mussel. 10. Pond bat. 11. Pearl mussel. 12. Pond bat. 13. Pond bat 14. Pond bat. 15. Pond bat. 16. Thick-shelled mussel 17. Pearl mussel. 18. Pond bat. 19. Pond bat. ● Habitats of the seasonal concentrations of birds.

ALIEN SPECIES

Several alien species occur in the Russian section of the Gulf of Finland basin. Some of them moved from the inland waters of Russia. A bivalve mollusc zebra mussel, *Dreissena poymorpha*, is especially well known. Its native distribution area is located in Caspian Sea basin, but it became widely distributed over last decades in Europe and Northern America. Zebra mussels often populate tubes and other constructions damaging the water flow. These mussels are especially harmful to the electric stations. Gulf of Finland is close to the northern borderline of the distribution area of *Dreissena*, that is why this threat was considered insignificant. However in 2013 such case was noted: zebra mussels (together with other small invertebrates) rapidly increased in numbers in canals at the southern coast of the Gulf of Finland (Popov, unpublished data) (Figure 12). Taking into account recent global warming, it is probable, that this invasion will progress, and *Dreissena* will populate even more northern water bodies.

A small fish Chinese sleeper, *Percottus glenii*, has been released into the Gulf of Finland close to Saint-Petersburg city about 100 years ago. It was brought from Amur basin as an ornamental fish (Reshetnikov, 2009). Several decades they remained unnoticed in new habitats, but now they became abundant. Chinese sleepers are especially noticeable in small water bodies, which are hardly suitable for the other fishes (Figure 11). These new inhabitants usually exterminate all small animals including the larvae of amphibians. In the big water bodies Chinese sleepers are not numerous, however they steadily settled in the coastal waters of the eastern extreme of the Gulf of Finland. Recently they had been found close to the Russian borderline at the southern coast of the Gulf of Finland. Their following spread is probable.

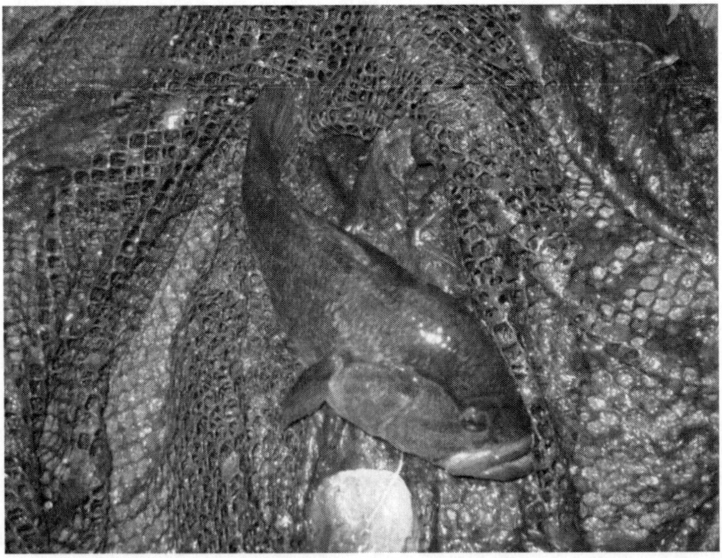

Figure 11. Chinese sleeper from a pond at the southern perimeter of Saint-Petersburg city.

An American small catfish - Brown bullhead, *Ameiurus nebulosus* – penetrated recently some lakes at the northern perimeter of Saint-Petersburg city. This invasion seems to be progressing (Litvinov, 2015). The role of this species in the local ecosystems is not studied yet.

The attempts to introduce new valuable fishes in local water bodies had been undertaken in the past several times, but unsuccessfully. Yet in XVIIIth century Russian Empress ordered to release sterlets, *Acipenser ruthenus*, at the Saint-Petersburg city. 18 000 sterlets had been really released in the Neva river, and some restrictions on the catching of them had been declared (Danilevsky, 1875). However these fishes had been caught soon after without any recovery. In 1960-70s several thousands of non-native sturgeons (Siberian sturgeon, *Acipenser baerii baerii*; *Acipenser baerii baicalensis* and Russian sturgeon, *Acipenser gueldenstaedtii*) (Berdichevsky et al., 1983) and Siberian white fishes had been released there, but no cases of their natural reproduction are known. Probably, only *Coregonus peled* settled in some lakes.

An alien crayfish species – narrow clawed crayfish, *Astacus leptodactylus* – is actively spreading over local water bodies. Its native habitats are located in Black and Caspian Seas basins. When it turns out to be in the same water body with noble crayfish, the last one disappears soon after. The mechanism of such a competition is not well known. Narrow-clawed crayfishes have been noted recently in the north-western part of the Russian section of the Gulf of Finland basin (Figure 12). Following spread is probable.

Figure 12. Narrow-clawed crayfish, Lake Lakhtinsky razliv (at the northern coast of the Gulf of Finland).

Figure 13. Marsh frogs captured during sampling of *Dreissena* at the southern coast of the Gulf of Finland.

One alien amphibian species was found at the Saint-Petersburg – Eurasian Marsh Frog, *Rana ridibunda*. Its main distribution area is located in more southern territories. Most likely, it does not exert significant influence on local ecosystems. *Rana ridibunda* populates mainly the coastal wetlands of the Gulf of Finland, where local amphibians are almost absent. Up to recent time they have been noted just at the eastern extreme of the Gulf, but now they are approaching the state borderline (they have been found in process of study of zebra mussels) (Figure 13).

As for the land alien species, some of them had been settled in the Gulf of Finland basin long ago because on intentional releases – muskrat, racoon dog, *Nyctereutes procyonoides*, American mink (Figure 14, 15). The activity on such releases had been stopped long ago. Among plant species the Sosnowsky's hogweed, *Heracléum sosnówskyi*, is the most noticeable one. Its number continuously increases (Figure 16). Sometimes it occupies the plots of several hectares.

Thus, Russian section of the Baltic Sea basin serves as a gate for the distribution of some species. The opposite process, i. e. the penetration of alien species from the west is also known. Recently European anchovy, *Engraulus encrasicolus*, reached Russian waters (Popov, 2014). Canadian beavers,

Castor canadensis, (Parker et al., 2012) and probably American crayfish, *Pacifastacus leniusculus*, penetrated Russia from European countries (Kouba et al., 2014).

Figure 14. Racoon dog, northern coast of the Gulf of Finland.

Figure 15. American mink, a forest in the north-western part of the Russian section of the Gulf of Finland basin.

Figure 16. Sosnowsky's hogweed at the south-eastern perimeter of Saint-Petersburg city.

CONCLUSION

The most remarkable feature of the Russian section of the Baltic Sea is a conservation of endangered habitats, which are continuously declining in Europe. Due to particularities of land use in Russia (absence of private property on rivers and forests) the refuges for wildlife often persist there even without special conservation measures in spite of high density of human population.

ACKNOWLEDGMENTS

The author is grateful to Alexander Marunchak and Dmitry Starikov for the pictures and assistance in field studies.

References

Aksenov, D; Dobrynin, D; Filonenko, I; Esipova, E; Kirillov, A; Kobyakov, K; Koltsov, D; Popov, I; Tokarev, P. Methods for delineation of the high conservation value. In: Atlas of high conservation value areas, and analysis of gaps and representativeness of the protected area network in northwest Russia Arkhangelsk, Vologda, Leningrad, and Murmansk Regions, Republic of Karelia, and City of St. Petersburg. Helsinki: Edita Prima Oy; 2013; 63-125.

Berdichevsky, LS; Maliutin, VS; Smolianov, II; Sokolov, LI; Akimova, NV; Kokarev, AV. In: Itogi rybovodno-acclimatizatsionnikh rabot s sibirskim osetrom. Biologicheskiye osnovy osetrovodstva. Moscow: Nauka; 1983; 259-270 (In Russian. "Results of hatchery-acclimatization works with Siberian sturgeon." Бердичевский Л. С., Малютин В. С., Смольянов И. И., Соколов Л. И., Акимова Н. В., Кокарев А. В. Итоги рыбоводно-акклиматизационных работ с сибирским осетром//Биологические основы осетроводства. М.: Наука. -1983. С. 259 - 270).

Berg, LC. O neobkhodimosti okhrany osetra v basseine Nevy. Za rubnuyu industriyu Severa, 1935, 7, 30-31 (In Russian. "About the necessity of protection of sturgeons in Neva basin." Берг Л. С. О необходимости охраны осетра в бассейне Невы. За рыбную индустрию Севера. 1935. № 7. С. 30-31)

Danilevsky, NY. Opisaniye rybolovstva v severo-zapadnikh oziorakh. Issledovaniye o sostoyanii rybolovstva v Rossii. T. IX. Saint-Petersburg: Ministerstvo gosudarstvennikh imushestv; 1875. (In Russian. "Description of fisheries in north-western Lakes." Данилевский Н. Я. Описание рыболовства в северо-западных озёрах. Исследование о состоянии рыболовства в России. Т. IX. СПб: Министерство государственных имуществ, 1875. 157 с).

Gaginskaya, AR. Mammals. In: Gogland Island. Results of the survey of the natural complexes. Saint-Petersburg: Tuskarora; 2006; 27-28.

Grimm, OA. Ryby i rybolovstvo v reke Luge. Selskoye khoziaistvo i lesovodstvo, 1889, CLXII, 121-139. (In Russian. "Fishes and fisheries in the Luga River." Гримм О. А. Рыбы и рыболовство в реке Луге// Сельское хозяйство и лесоводство. 1889. Ч. CLXII. С. 121–139).

Edsman, L; Füreder, L; Gherardi, F; Souty-Grosset, C. Astacus astacus. The IUCN Red List of Threatened Species 2010: e.T2191A9338388. http://dx.doi.org/10.2305/IUCN.UK.2010-3.RLTS.T2191A9338388.en. Downloaded on 02 January 2016.

Kazakov, R. Condition of fish stock, yield to fishery and migrations of Atlantic salmon from rivers of the USSR to the Baltic Sea. *Finnish Fisheries Research*, 1985, 6, 43-54.

Kouba, A; Petrusek, A; Kozák, P. Continental-wide distribution of crayfish species in Europe: update and maps. *Knowledge and Management of Aquatic Ecosystems,* 2014, 413, 05, 1-31.

Kuziakin, AP. Letuchie myshi. Moscow: Sovetskaya nauka; 1950. (In Russian. “Bats.” Кузякин А. П. Летучие мыши. М.: Советская наука, 1950. 443 с).

Litvinov A. Report on amiur fishing. http://www.youtube.com/watch?v= GMzRZrg55vQ. Last accessed 1 December, 2015.

Nezhikhovsky, RA. Reka Neva i Nevskaya guba. Gidrometeoizdat: Leningrad; 1981. (In Russian. “Neva River and Neva Bay.” Нежиховский Р. А. Река Нева и Невская губа Ленинград Гидрометеоиздат 1981).

Noskov, GA. The main results of bird migration studies in the North-West Region of Russia. In: Study of the Status and Trends of Migratory Bird Populations in Russia. Saint-Petersburg: World and Family; 2002; 62 -79.

Ostrovsky, A; Popov, I. Freshwater pearl mussel Margaritana margaritifera (Unionidae, Margaritinidae) in the rivers of the reserved area “Gladyshevsky” (Leningrad oblast). *Russian journal of zoology*, 2008, 87, 5, 1-2. (In Russian with English summary).

Parker, H; Nummi, P; Hartman, G; Rosell F. Invasive North American beaver *Castor canadensis* in Eurasia: a review of potential consequences and a strategy for eradication. *Wildlife Biology*, 2012, 18, 4, 354-365.

Popov, IYu. Salmon fishes of the river Luga (Sanct-Petersburg region) International conference. Biodiversity of the European North. The Book of Abstracts. Petrozavodsk: RAS; 2001; 302.

Popov, I. The growth character of the Baltic salmon (*Salmo salar*) of the Russian part of the Gulf of Finland of the Baltic Sea. Problems of fisheries, 2010, 11, 3 (43), 415-427. (In Russian with English summary).

Popov I. New Fish Species in the Russian Part of the Gulf of Finland and Inland Water Bodies of St. Petersburg and Leningrad Oblast. *Russian journal of biological invasions,* 2014, 5, 2, 90-98.

Popov, I. Activités de conservation sur la zone située autour de Saint-Pétersbourg et leur impact sur la mulette perlière. *Penn ar Bed*, 2015, 221, 22-27.

Popov, IYu; Ostrovsky AN. Survival and extinction of the southern populations of freshwater pearl mussel *Margaritifera margaritifera* in

Russia (Leningradskaya and Novgorodskaya oblast). *Hydrobiologia,* 2014, 735, 1, 161-177.

Pravdin, IF. Osetroviye ryby v Ladozhskom i Onezhskom ozerakh. Izv. Karelo-finskoy nauchnj-issledovatelskoy basy AN SSSR; 1948; 4; 73-79. (Pravdin I. "Sturgeons in Ladoga and Onega Lakes" Правдин И. Ф. Осетровые рыбы в Ладожском и Онежском озерах//Известия карело-финской научно-исследовательской базы АН СССР. №4. Петрозаводск. 1948. С. 73-79).

Red book of Russia. http://www.sevin.ru/redbook Last accessed 1/12/2015 (Official web-page of the Russian Red book).

Reshetnikov, AN. Current area of Amur sleeper (*Perccottus glenii* Dibowski, 1877) in Eurasia. *Ross. Zh. Biol. Invazii,* 2009, 1, 17–27.

Skumatov, DV. Evropeiskaya norka v Rossii: sovremennoye sostoianiye i perspektivy sokhraneniya v usloviiakh okhotnichiego promysla. PhD Thesis. Viatskaya selkhoz. acad. 2006. (In Russian. "European mink in Russia: current condition and perspectives of conservation in conditions of commercial hunting." Скуматов Д. В. Европейская норка в России: Современное состояние и перспективы сохранения в условиях охотничьего промысла. Дисс. канд. н. 2006. Вятск. Сельхоз. акад. 159 с).

Thielen, F. (Editor). Rearing of unionoid mussels (with special emphasis on the Freshwater Pearl Mussel *Margaritifera margaritifera*). Ferrantia. 2011. T. 64. Musée national d'histoire naturelle, Luxembourg, 66 p.

Water code of Russia. (03. 02. 2006. № 74-FZ www.pravo.gov.ru).

Zhadin, VI. Fauna SSSR. Molliuski. T. IV. Moscow, Leningrad: AN SSSR; 1938. (In Russian. "Fauna of USSR. Molluscs." Жадин В. И. Фауна СССР. Моллюски. Т. IV. Вып. 1. Сем. Unionidae. М. Л.: АН СССР, 1938. 170 с).

In: Baltic and Black Sea
Editor: Andre Mills
ISBN: 978-1-63485-233-3

Chapter 2

ICHTHYOPLANKTON BIODIVERSITY IN THE DEEP-WATER AND COASTAL PARTS OF THE SOUTH-EASTERN BALTIC. CONSEQUENCES OF ENVIRONMENTAL CHANGES

E. M. Karasiova[*]
AtlantNIRO, Kaliningrad, Russian Federation

ABSTRACT

The Baltic Sea could be considered as an example of the successful ecosystem functioning under the small biodiversity of biota. The long-term data on the ichthyoplankton species diversity of the south-eastern Baltic gives an opportunity to examine its dynamics on spatio-temporal aspects. The deep-water part and shallow zone of the Baltic Sea are sharply distinguished by species composition, abundance and ecological adaptations of egg and larvae assemblages. Only 4 fish species with pelagic eggs (cod, sprat, flounder, rockling) spawn in the deep basins of the Eastern Baltic. The scarcity of species richness could be compensated by specific adaptations: the extremely prolonged spawning season (cod, sprat, rockling), a transition to different types of spawning biotopes in the course of reproductive season (sprat) or in

[*] E-mail: karasiova@rambler.ru.

different areas of the Baltic (flounder). Under the invariable species richness through several decades, the Shannon's index of general diversity demonstrated a tendency to decline with minimum in the late 1990s - early 2000s due to dramatic environmental changes. Currently, early developmental stages of sprat occupy the dominant position in the deep-water ichthyoplankton assemblage of the eastern Baltic. The spawning habitats in the coastal shallow zone are used mainly by fishes laying the bottom eggs. Larval assemblages included 22 species. This increase of species richness was defined by the bottom substrate diversity, the seasonality of temperature regime, depth gradients and alternations of upwelling and downwelling processes. Long - term changeability of species richness could be traced between 1946-1955 and 1992-2007 based on literature sources and field investigations. In the coastal zone a trend was also observed to decline of general diversity indices. The one of tendencies was a diminution of occurrence and habitat fragmentation of larval species from winter spawning (snakefish, rock eel) on contrary with dominance of fish larvae from summer reproduction (gobies). A mutual trend was revealed to transition from polydominant to monodominant assemblages both in deepwater basins and shallow coastal zone. On the open part of the Baltic this was linked to abundance of early developing stages of small pelagic sprat while in the coastal zone with dominance of small short-lived bottom species, a namely, sand goby. These tendencies were linked to the climatic and environmental changes during last decades.

Keywords*:* deep-water and coastal ichthyoplankton, long-term trends of diversity changes

It is well-known that all self-organizing living systems need the minimal species diversity in order to provide for the cyclic redistribution of energy flows between components of ecosystems (Odum, 1971). The strategy of nature is to increase biodiversity, but only up to limits which ones do not lead to the drop of the resource use effectiveness.

The Baltic Sea is an example of successful ecosystem functioning under the small diversity of biota (Elmgren, Hill, 1997). The reasons of this phenomenon have their roots in geological youth of this water body and unique peculiarities of abiotic environment. For the last decades of 20th century the Baltic Sea ecosystem has undergone significant changes. The replacing of dominant species in zooplankton and ichthyofauna also is among their changes (Moellman et al., 2009). The suitable subject for studying long-term responses of biodiversity on the ecosystem transformation is ichthyoplankton, i.e., fish eggs and larvae. This is determined by availability of long-term observations revealing spatial-temporal variability of abundance and species composition. The prolonged history of ichthyoplankton investigations in the Baltic Sea, the tradition of the primary data publication, similar catchability of sampling gears used (Oeberst at al., 1981) and finally beginning of the regular ichthyoplankton surveys in the USSR and Germany have given possibilities to receive the bulk of representative data for last 50-70 years.

Coming from the presence or lack of halocline, the basic feature of the Baltic hydrography, two main subsystems can be distinguished in this ecosystem:1) deep-water basins, 2) shallow coastal zone. Analogously, there the marine and coastal ichthyoplankton assemblagies can be revealed. The first of them includes pelagic eggs and larvae of pelagophilous fishes, while the larvae of fishes spawning bottom eggs on underwater substrates are related to second one.

The aim of this study is revealing spatio-temporal dynamics of ichthyoplankton biodiversity in different biotopes of the Baltic Sea.

Materials and Methods

Long-term data on the ichthyoplankton species composition, abundance and distribution were used on the basis of literary sources (Mankowski, 1950, 1951, 1955, 1959; Lindblom, 1973; Grauman, 1980, Grauman, Lisheva, 1990; Karasiova, Voss, 1994, Karasiova, Ivanovich, 2007; Karasiova et al. 2011) as well as the AtlantNIRO Baltic Sea Lab dataset. Currently ichthyoplankton surveys were carried out by the Baltic Sea lab since 1992 to 2015. Ichthyoplankton nets, Hensen net in the Polish sampling in1940-1950s and IKS-80 net in the Soviet and Russian sampling in1950-2000s, were used. Vertical hauls were conducted in layer from bottom to surface. Simultaneously with vertical sampling there were made 5-minute horizontal subsurface hauls in coastal zone in depth range of 05 - 60m in July1999 - 2002 and in depth range of 1.0-1.5m in summer 1999 and 2014. Quantity of current samples used in this study was as follow: 1) 262 samples in deep basins during spring-summer 2003 and 2006 including transects through the Baltic Sea in 2006; 2) 170 inshore horizontal hauls in 1999-2002, 2014 and 562 vertical samples in 1992-2007. The abundance was estimated as sp./m2.

The total number of fish egg species in deep-water basins and larval species number in coastal zone were used as the indices of species richness S.

The Shannon's diversity index Hs was applied to characterize the general species diversity: Hs = -∑ (ni/N) (log2 (ni/N), where ni – number of each species, N –total abundance. The Simpson's domination index C was used to characterize the domination degree in ichthyoplankton assemblies: C = ∑ (ni/N)2. The Pielow's equitability index E was estimated as E= Hs/log S, where Hs = Shannon's diversity index, S – species richness. The above parameters were estimated according Fedorov, Gilmanov, 1980. These calculations were based on mean fish egg values by areas and seasons from above literary sources and AtlantNIRO data.

The data on hydrografic conditions and wind regime were obtained from the AtlantNIRO dataset and literary source (Zezera, 2002). The frequency (W, %) of north-easterlies, westerlies and southerlies was estimated for time intervals included month before the July survey to its last day. The locations of the Baltic deep-water basins (Bornholm and Gotland Basins, Gdansk Deep) as well as the coastal zone of the Kaliningrad region with adjacent areas were shown at Figure 1.

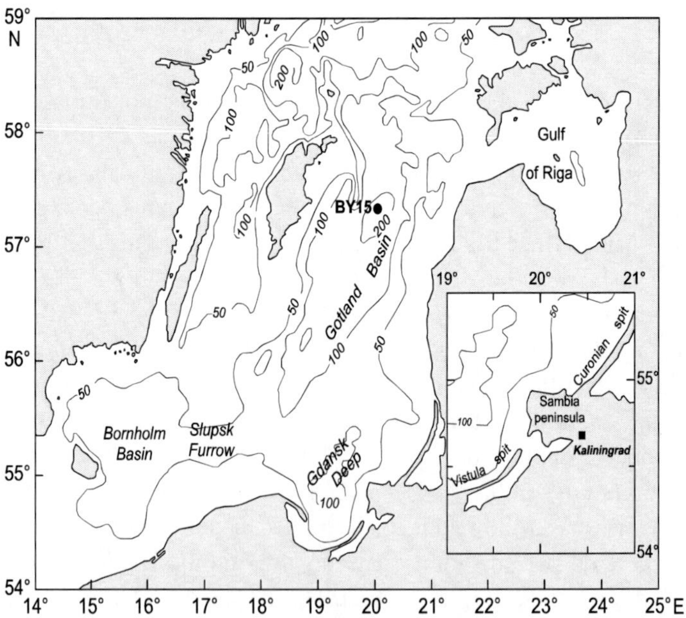

Figure 1. The map of the Baltic proper and the Kaliningrad region location.

RESULTS

1. Deep-Water Basins: Long-Term Changes in Environmental Conditions and Ichthyoplankton Diversity Variability

The assemblage of pelagophilous fishes, reproducing in the Baltic deepwater basins, was formed under the environmental influences, which extremely have limited its species richness. The low salinity is one of the most important among those factors. The centennial maximum of bottom salinity in the central part of Gotland Basin (Station BY15) observed in 1953, amounted to 13.8 psu at depth 200 m (Soskin, 1963). Accordingly, the highest decadal bottom salinity was estimated for 1951-1960 (Figure 2). Afterwards, in 1983 the prolonged lack of major inflows caused the drastic decline of bottom salinity in so-called Eastern Basins: Gdansk Deep, Southern and Central parts of Gotland Basin to a minimum in 1992-1993. According to the characterization of the Baltic hydrographic regimes the Baltic Sea attained the contemporary ecological status since the regime shift of 1988-1992 (Alheit et al., 2005). That shift was also linked to the strengthening of NAO activity and increasing water temperature (BACC Author Team, 2008). The thermal regime change in the SE Baltic could be confirmed by decadal air temperature dynamics during winter-spring season in Kaliningrad region with positive trend for 1951-2010 (see Figure 2).

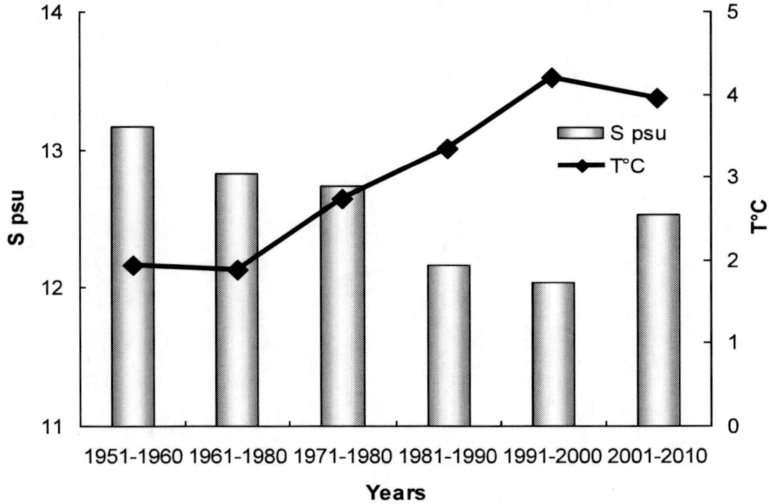

Figure 2. The variability of bottom salinity S and air T°C in winter-spring in Kaliningrad region.

Transformation of the deep-water environment has consequences for the ichthyoplankton assembly diversity.

The ichthyoplankton species composition in the deep-water basins includes the eggs of four pelagofilous fish species:1) cod Gadus morhua callarias L., 2) flounder Platichthys flesus L.; 3) rockling Enchelyopus cimbrius L.; 4) sprat Sprattus sprattus balticus (Schneider). While flounder eggs occurred from February to early June, another egg species could be found from February to August (sprat), from February to October (rockling), and from late January to October (cod).

Some specific features of the species egg distribution in 1953-1954, 1970-1971 and 2003 - 2006 can be clearly displayed by cross-sections through the Baltic Sea (Figure 3-5). The primary data of Mankowski (1959) and Lindblom (1973) were used for Figure 3 and 4.

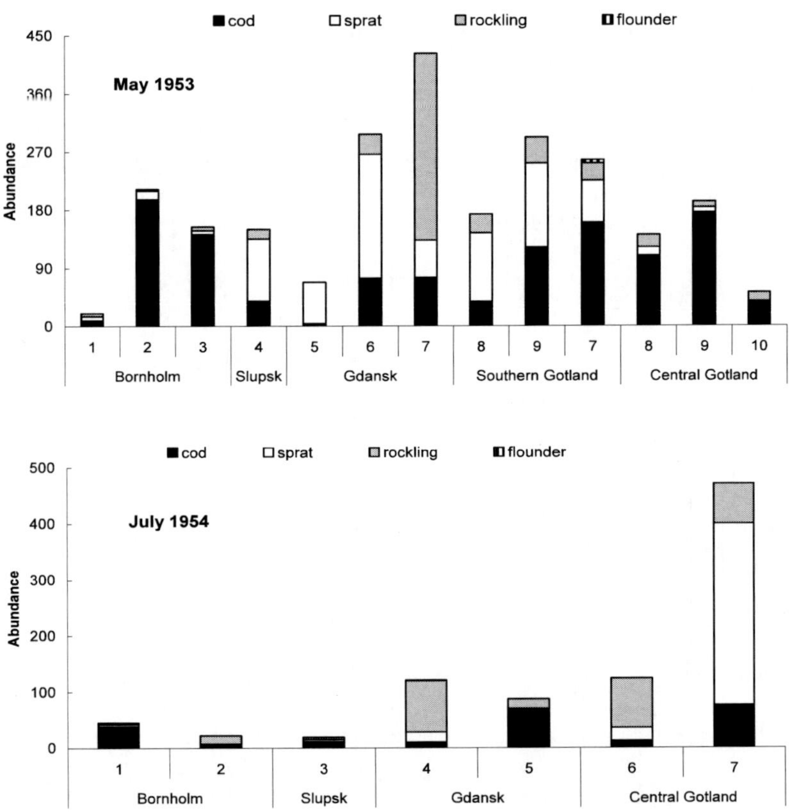

Figure 3. Spatial and seasonal distribution of fish eggs in 1953-1954 along transect through deep basins.

Cod eggs were abundant in all above mentioned deep basins under spring peak timing during the first period. In May 1953 cod eggs dominated in Bornholm and Central Gotland Basins. The quantities of cod and sprat eggs were almost equal in Southern Gotland. Gdansk Deep was dominated by sprat eggs. During the summer spawning phase (July 1954) cod eggs kept the dominance in Bornholm Basin as well sprat eggs in Gdansk Deep under the noticeable abundance diminution. There was a displacement of dominant egg species from cod to sprat in the Central Gotland Basin under the significant sprat egg number increasing.

Seventeen years later cod eggs extended in Southern and Central Gotland Basin in May-June 1970, being the subdominant species in those areas with dominance only in Bornholm Basin, according Lindblom data, 1973 (Figure 4). Sprat eggs prevailed in May in Gotland Basin; prevalence of eggs of that species was also obvious in summer 1971. Cod eggs dominated in Bornholm Basin only in February and August in average for 1968-1977 (Grauman, 1980). Actually sprat eggs often occurred in dominant position in 1970s in all deep-water areas.

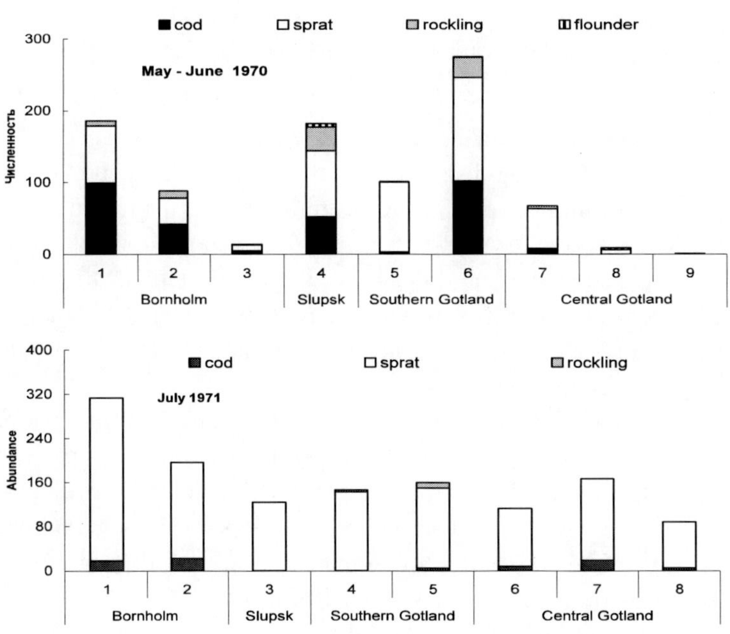

Figure 4. Spatial and seasonal distribution of fish eggs in 1970-1971 along transect through deep basins.

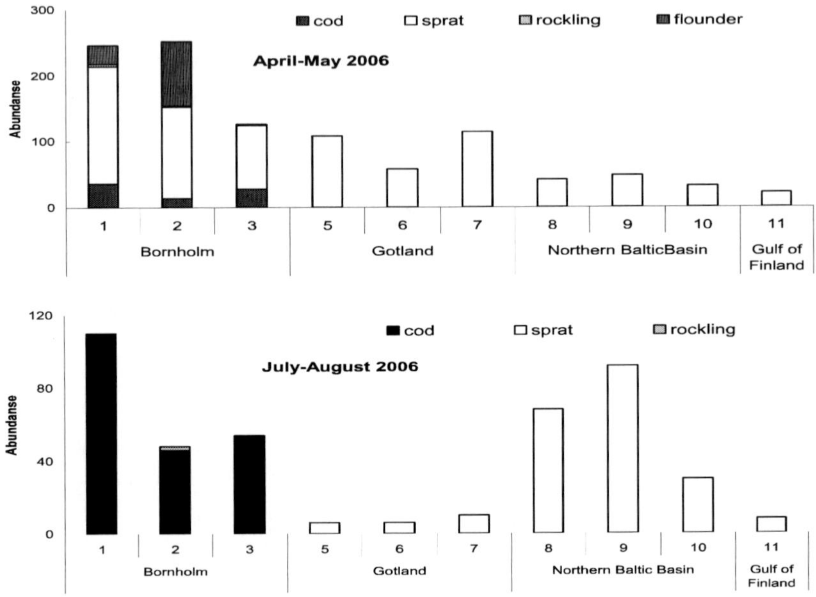

Figure 5. Spatial and seasonal distribution of fish eggs in 2006 along transect through deep basins.

The current status could be demonstrated by the cross-section via the Baltic Sea in 2006. Cod egg distribution in that year was limited by Bornholm Basin only. Due to shift of cod spawning timing from spring to summer its egg abundance was very low in April-May. Sprat occupied the dominance position while flounder was at the sub-dominance one. In Eastern Basins sprat also significantly contributed in fish spawning activity. Cod egg abundance sharply enlarged in August in Bornholm Basin on background of the very insignificant the rockling egg occurrence.

Results of estimations of Shannon'a indices for different time periods for spring, summer separately as well as the combined ones are presented in Tables 1-3. Combining Shannon's indices for the entire spawning season could increase as compared with indices for separated months. Such the increase took place in cases of dominant species displacement. The enlargement was observed in the Central Gotland in 1953 - 1954 due to the dominance shift from cod to sprat. While Shannon's index was equal 0.793 in spring and 1.370 in summer, it grew to 1.529 for the entire spawning season. The same situation was revealed also in Gdansk Deep in 1953-1954 (seasonal shift in dominating from sprat to rockling). A relatively low Shannon's index in Bornholm Basin was determined by overwhelming cod egg dominance and lack of sub-dominant species. Probably it was a seldom situation being the specific trait for those years.

Table 1. The dominant species and Shannon's diversity index Hs in deep basins for 1953-1954

Month	Variables	Basins				
		Bornholm	Gdansk	S. Gotland	C. Gotland	All basins
V	Dominant species	Cod	Sprat	Cod, sprat	Cod	Cod
	H_s	0.589	1.348	1.484	0.793	1.363
VII	Dominant species	Cod	Rockling	Sprat	Sprat,	Sprat
	H_s	1.496	1.322	1.324	1.370	1.492
V-VII	H_s Dominant species	Cod	Sprat	Sprat	Sprat	Cod, sprat
	H_s	0.791	1.580	1.480	1.529	1.542

There was no seasonal dominating displacement for 1968 – 1977; therefore combined seasonally Shannon's indices were on average level between spring and summer values.

Table 2. The dominant species and Shannon's diversity index Hs in deep basins for 1968-1977

Month	Variables	Basins				
		Bornholm	Gdansk	S. Gotland	C. Gotland	All basins
III-V	Dominant species	Sprat	Sprat	Sprat	Sprat	Sprat
	H_s	1.707	0.887	1.306	1.491	1.362
VI-VIII	Dominant species	Sprat	Sprat	Sprat	Sprat	Sprat
	H_s	0.835	0.722	0.579	0.928	0.767
VI-VIII	Dominant species	Sprat	Sprat	Sprat	Sprat	Sprat
	H_s	1.202	0.825	0.868	1.213	1.053

The diversity level in Bornholm Basin in August 2006 seems to be too low (0.062). However, it will rise to 1.305 when combining data for the entire season. This was linked to the dominance shift from sprat to cod. Thus, the seasonal dismatch in spawning timing of key fish species led to increase of the combined diversity index.

Table 3. The dominant species and Shannon's diversity index Hs in deep basins for 2003, 2006

Month	Variables	Basins				
		Bornholm*	Gdansk	S. Gotland	C. Gotland	All basins
V	Dominant species	Sprat	Sprat	Sprat	Sprat	Sprat
	H_s	0.930	0.196	0.113	0.057	0.423
VII-VIII	Dominant species	Cod	Sprat	Sprat	Sprat	Sprat
	H_s	0.062	0.285	0.299	0.134	0.829
V-VIII	Dominant species	Sprat	Sprat	Sprat	Sprat	Sprat
	H_s	1.305	0.242	0.179	0.081	0.627

* Bornholm data for IV-V and VIII 2006.

The long-term dynamics of Shannon's indices from1950s to 2000s was characterized by decline in all Eastern Basins for all seasons to the exclusion of Bornholm Basin (Figure6). The last one was linked at first with mono-dominant species composition in Bornholm area for 1953-1954, in contrast with appearance of subdominants later in this area. The most contribution into the drop of diversity for the last six decades took place during the summer spawning time. Although the major inflow 2003 improved environmental conditions for cod reproduction there was not a prolonged intensification of spawning in eastern Basins next years after summer 2003.

The combined Shannon's index estimated for entire season and areas have demonstrated the drop from 1.54 in 1953-1954 through 1.05 in 1968-1977 to 0.63 in 2003-2006. This change was resulted in the transition from polydominant to monodominant fish community due to the drastic decline of the Eastern Baltic cod stock (Koester et al., 2003).

It is worthy noting that the extremely low taxonomic ichthyoplankton species richness in deep-water basins of Baltic proper can be compensated to a certain degree by other diversity manifestations. They include the very prolonged spawning season and shifting in spawning peak timing under deteriorating ambient conditions for one of key species. Among other compensation responses it could be shown out the changing of sprat spawning biotope from the upper part of halocline in spring to the surface thermocline in summer. It is well-known that flounder in the Baltic used different spawning biotopes in the large spatial scale. This species formed two ecological groups, one of which reproduces in deep-water basins spawning pelagic eggs under relatively high salinity provided their buoyancy. Another ecological group of flounder reproduces on bottom substrates of shallow banks in the NE Baltic at lower bottom salinity (Florin, Hoeglund, 2008).

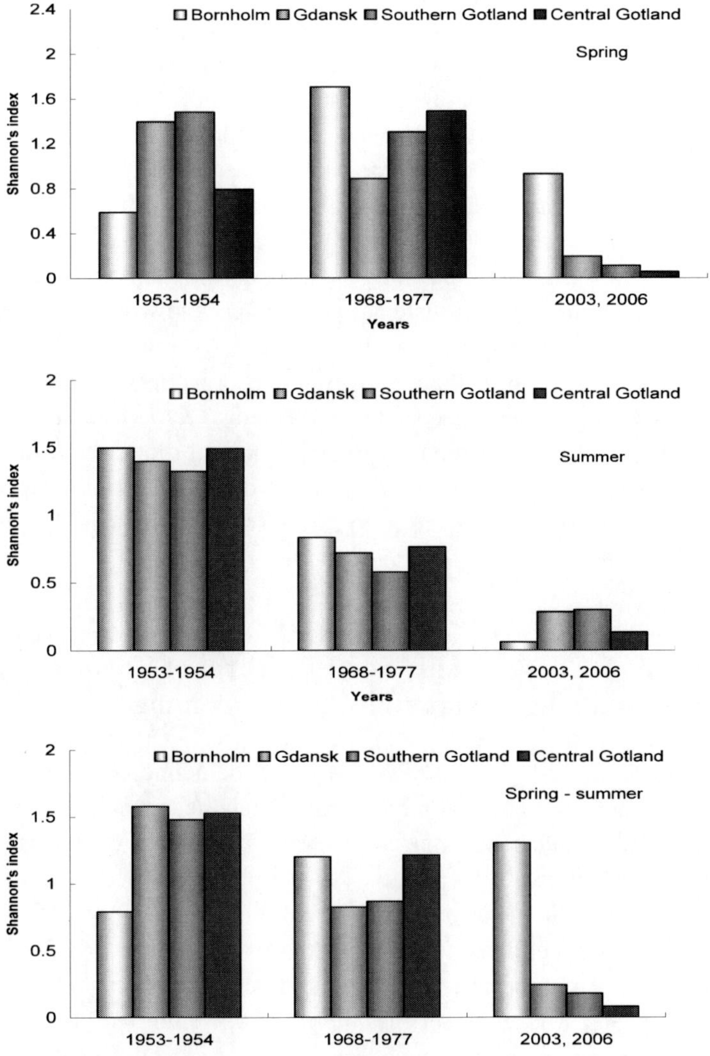

Figure 6. The inter-annual and seasonal variability of Shannon's indices in deep basins.

The time periods under consideration demonstrated three different situations in the spatio-temporal dynamics of reproductive activities of key fish species. In early 1950s the cod peak spawning timing was in April - May. The sprat peak spawning timing was two months later in Gotland Basin, namely, in July. The earlier spawning provided better survival of cod progeny. Thus, cod received a competitive advantage in1950s.

Later, in 1970s, cod spawned earlier than sprat, as before. However the difference in peak spawning diminished. Mass cod spawning took place in Bornholm Basin in April-July; in Gdansk Deep and Gotland Basin in May-July and April-June correspondingly (Grauman, 1980). Sprat peak spawning timing was a few later: in May-July in Bornholm Basin and June-July in Gotland Basin. In 1970s cod could receive a competitive advantage during the major inflows of Kattegat water in the Baltic while warm winters provided the advantage for sprat.

Currently the sprat mass spawning was usually in late-April – early July; spawning peak timing shifted to late May-July after cold winters. On contrast, the mass spawning of cod shifted to July - August. As result it led to later cod larvae hatching, reduction of their growth period in warm season and change of species composition and size of the food items (Karasiova et al., 2010). Correspondingly, chances of the cod progeny survival diminished. In the end the late cod spawning led to a decline of productivity of their stock. The competitive advantage of sprat linked to earlier reproduction could be realized in years with hydrographic regime favorable for its offspring.

2. The Shallow Coastal Zone: Some Features of Environmental Conditions and Ichthyoplankton Diversity Changes

The coastal zone is characterized by the halocline absence and low salinity. The mean long-term salinity in coastal stations in the Kaliningrad area constituted 5.5 psu being in range from 4.1 to 7.1 psu. In contrast there the seasonal water temperature variability was sharply pronounced. According to data for 2000-2004 the mean water temperature could amount to 16.5-16.7°C in summer while in February water temperature could drop to 1.5-1.9°C. At the same time, even in the warmest months of year, the water intrusions from the cold intermediate layer under the impact of upwelling are able to reduce the near-bottom water temperature. Taking into account the alternation of upwelling - downwelling events and wind - inducing waving, the near-shore shallow areas provide much more diversified but less stable conditions as compared to the deep-water basins.

The species richness of coastal larval assembly is noticeably higher than that of open sea depths. Some groups factors contribute to increasing species number in the coastal zone: 1) seasonal thermal regime variability; 2) bottom substrate diversity (sand, gravel, pebbles, boulders, silt, shells, seaweeds; 3) depth gradient (from 0.5 to 60 m accompanied by ambient condition changes, 4) upwelling - downwelling alternations. Due to the first factor the assembly can be divided on larvae from different seasonal origin, e.g., winter (rock eel), spring (spring-spawning herring), summer (gobies) and autumn (autumn-spawning herring). Considerable seasonal variability of the water temperature facilitates differentiation of spawning biotopes temporally and thus provides more complete utilization of resources of coastal habitats.

Due to next two factors there were lithophils (bullrout, see scorpion), phytophils (garfish, sea snail), psammophils (greater and lesser sand eel). The sand goby often lay eggs on shells in mussel biocenosises. Its depth range is more expanded as compared with brackish-water common goby. The wind regime and upwelling - downwelling events influence larval abundance and species richness. As result, the coastal zone provides a mosaic of different habitats for species with bottom eggs.

The list of species composition includes 22 species of larvae (Table 4). The coastal assembly includes 17 species either having origin from fish laying the bottom eggs (15 species from Fam. Ammodytidae, Belonidae, Clupeidae, Cottidae, Gasterosteidae, Gobiidae, Liparidae, Pholididae, Scophthalmidae) or bearing eggs in brood camera (2 species from Fam. Syngnatidae). Recently a number of Gobiidae species increased due to invasion of the ponto-caspian round goby. However, the abundance of the most species above mentioned was low amounted usually 1-3 sp./5-minute haul.

Besides, larvae of pelagofilous fishes can occur in the coastal zone: sprat, flounder, rockling, cod and very seldom, plaice. Sprat egg also can be distributed over depths less 50 m, especially in summer season.

The seasonal variability of species richness demonstrated the maximum in July, coinciding with results of the Polish investigations in 1947-1955 (Figure 7).

In addition to Polish data, currently in July there were larvae of turbot, stickleback, sea scorpion, sea snail, broad-nose pipe fish and round goby. The seasonal abundance dynamics was characterized by the gradual growth from late winter to summer and following decrease to autumn. The dominant species was sand goby, amounting to usually 70-90% in ichthyoplankton. The highest concentrations of its larvae were observed in the first decade of July, being a consequence of spawning in June. Therefore it can be supposed that environment conditions during above months are the most important for the larval goby survival in coastal habitats.

Table 4. List of fish larvae species in coastal zone of the SE Baltic Sea

Species composition	Some specific traits
Coastal Assembly	
1. Snakefish - *Lumpenus lampretaeformis*	Bottom eggs
2. Rockeel - *Pholis gunnellus*	-"-
3. Bullrout - Myoxocephalus scorpius	-"-
Coastal Assembly	
4. Srtipped sea snail - *Liparis liparis*	-"-
5. Long-spined sea scorpion - *Taurus bubalis*	-"-
6. Herring - *Clupea harengus membras*	-"-
7. Greater sand eel - *Hyperoplus lanceolatus*	-"-
8. Lesser sand eel – *Ammodytes tobianus*	-"-
9. Stickleback – *Gasterosteus aculeatus*	-"-
10. Garfish – *Belone belone*	-"-
11. Sand goby - *Pomatoschistus minutus*	-"-,
12. Common goby - *Pomatoschistus microps*	-"-
13. Black goby - *Gobius niger*	-"-
14. Round goby - *Neogobius melanostomus*	-"-
15. Broad-nosed pipe fish - *Syngnathus typhle*	Bearing eggs in brood camera
16. Strait-nosed pipe fish – *Nerophis ophidion*	-"-
17. Turbot - *Psetta maxima*	Pelagic eggs under high salinity
Marine assembly	
1. Sprat - *Sprattus sprattus balticus*	Pelagic eggs
2. Flounder – *Platichthys flesus*	-"-
3. Plaice – *Pleuronectes platessa*	-"-
4. Rockling – *Enchelyopus cimbrius*	-"-
5. Cod – *Gadus morhua*	-"-

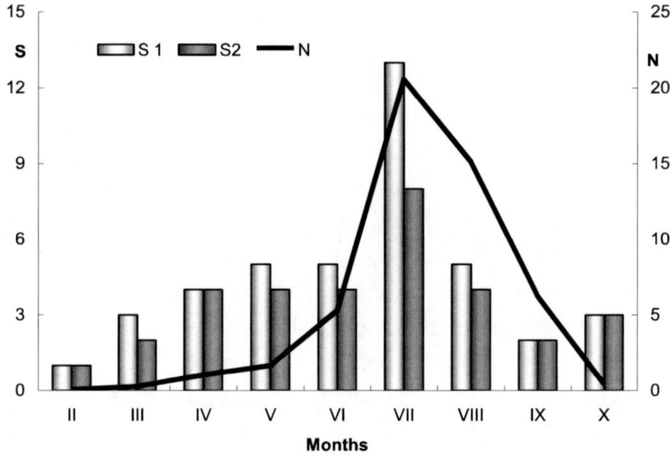

Figure 7. Seasonal dynamics of species richness S1 (Polish data), S2 (Russian data) and mean larval abundance (sp./m2).

As it is well-known the strong upwelling could result in the cold inflows from the cold intermediate layer to the coastal zone (Bychkova, Viktorov, 1987; Golenko, Golenko, 2012), which ones can block ripening and spawning of coastal fish on the significant part of their habitats. The upwelling formation along the eastern Baltic shores is linked to winds of northern and eastern directions while westerly cause downwelling event. For 1999-2002 the highest frequency of NE winds was observed in June-July 1999 (Table 5). Westerly prevailed in 2000 and 2001, while southerly showed the high occurrence in 2002.

Table 5. Wind frequency (%) and velocity in June - July 1999-2002

Wind directions	Wind frequency, %				Wind velocity, m/sec			
	1999	2000	2001	2002	1999	2000	2001	2002
N, NE, E, SE	53.88	23.09	30.77	9.38	2.56 1-6	1.75 1-2	1.85 1-2	2.0 -
W, NW	15.38	46.17	53.85	43.75	2.25 2-3	4.43 1-8	6.00 4-10	6.79 1-12
S, SW	30.73	30.73	15.38	46.88	3.62 2-5	4.42 1-8	2.0 -	4.79 2-7

The survey in July 1999 coincided with the late stage of upwelling development when the main feature was the inflow cold water into the coastal bottom layer. According to hydrographic situations mentioned above, the near-bottom water temperature distribution along the depths gradient was characterized either a sharp reduction from 20°C to 4°C (in 1999), or almost total absence of the temperature gradient at the mean near-bottom temperature of 15°C (in 2000). In 2000 the bottom temperature at depth range from 05 to 50 m decreased from 16°to 14°C. In 2001 and 2002 the situation was characterized with a weak temperature gradient within the depth range from 10 to 30 m and a more pronounced gradient at the depths 30-50 m. In these years cold water inflows like them of 1999 did not observed.

The most typical traits of larval distribution against water temperature can be traced at transects through the coastal zone along the depths gradient during July surveys (Figure 8). In July 1999 with very low abundance and species richness, the most number of gobies larvae concentrated near the shore at the depths less 10 m, while sprat larvae occurred only along the external boundary of the study area at the depths more than 40 m. In July 2000 with high abundance and species richness, the maximum number of gobies larvae was found at the depths from 20 to 25 m. Sprat larvae occurred in the whole depths range from 10 to 50 m. In July 2001 sprat eggs penetrated to the coastal zone to the depth of 10 m. In July 2002 with average levels of abundance and species richness the numbers of gobies larvae were caught much further offshore, at the depths from 30 to 50 m, as compared to the previous years. The distinguishing feature in 2002 was a mass appearance of lesser sand eel larvae with maximum to 50 sp./5-minute haul.

The variability in abundance of sand goby correlated with westerly frequencies (Figure 9a). The high bottom temperature was favourable to the outburst of its reproduction in 2000 and 2001. In contrast in 2002 the frequency increase of southerlies caused the water pile-up at the southern shore and, on contrary, water outflow from the northern one. As result the lesser sand eel spawning was intensified along the Vistula spit and the southern coast of Sambia peninsula, where sand grounds prevailed. The abundance of sand eel was positively linked to the difference of north-eastern and southern wind frequencies (Figure 9b).

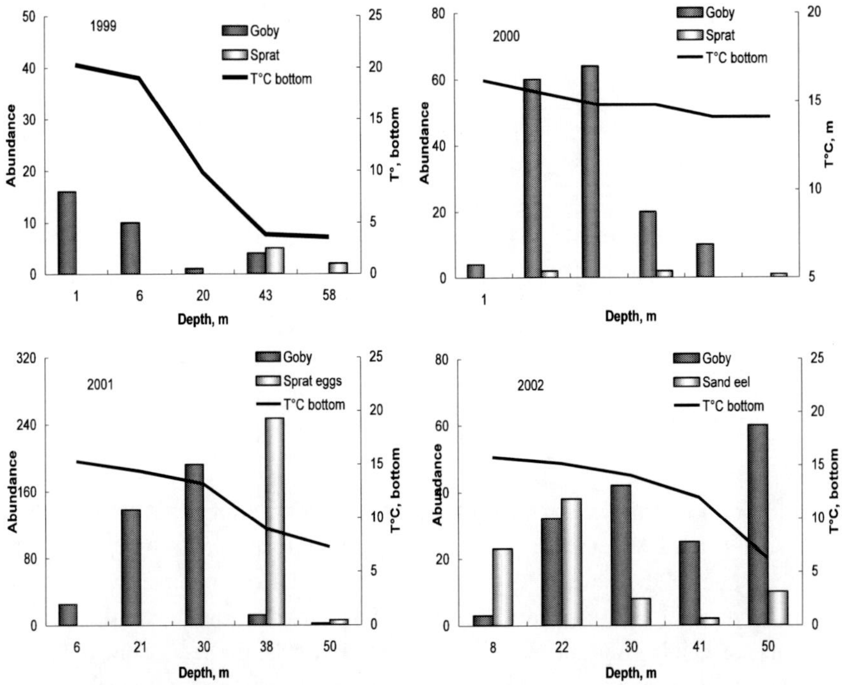

Figure 8. Larval abundance (sp./5 minute/haul) of key species and T°C bottom through cross-sections in coastal zone.

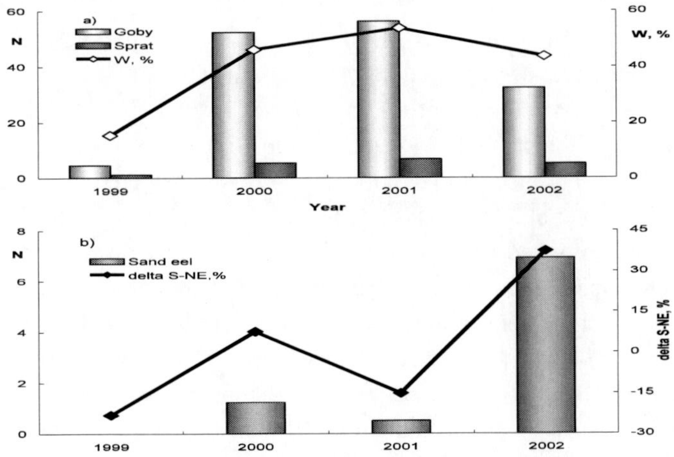

Figure 9. Larval abundance (sp./5min./haul) and wind frequency, %: a) goby and sprat larvae vs. westerlies; b) sand eel vs. difference between north-easterlies and southerlies.

Table 6. Shannon's diversity index Hs in larval assemblies of different species composition

Year	Depth strata, m	Coastal larvae	Coastal, marine larvae
1999	05-30	0.355	0.519
	05-60	0.288	1.097
2000	05-30	0.732	0.806
	05-60	0.589	0.951
2001	05-30	0.626	0.790
	05-60	0.637	0.998
2002	05-30	0.982	0.932
	05-60	0.824	1.242
1999-2002	05-30	0.800	0.932
	05-60	0.692	1.080

The Shannon's index for the coastal assembly was the least in 1999 and the maximal in 2002 (Table 6). These values displayed objectively the depressed state of the coastal assembly under impact of cold water inflow in 1999. The growth of that index in 2002 was a consequence of decreasing abundance of sand goby and enlargement of sand eel numbers. The including of marine larvae into calculations of H_s is evidently not admissible. The coastal zone is the natural habitat for post-metamorphosis sprat fry as well flounder. However the mass transportation of sprat larvae due to westerly strengthening in June-July usually led to poor recruitment (Karasiova, 2006). Ii is also known that the drift of cod larvae beyond the spawning site also resulted in unsuccessful generations (Hinrichsen et al, 2001).

There are some data about the role the sea stretch (1-2 m width), attached to sea beach, in the coastal larvae protection from negative upwelling impact. Due to the tide absence in the Baltic this site is considered as the upper sub-littoral (Paavola et al., 2005). The stable environment has arisen in that habitat owing to of prolonged calm weather during upwelling event. As result some coastal fish species could use that habitat for the reproduction. The low number of ichthyoplankton under upwelling event in the main habitats (05-30 m depth range) could partially be compensated by its reservation in the most water-shallow habitat. The evidence can be results of sampling (5-minute haul) in early July 1999 and late July 2014. In the first case the live eggs of garfish were found on seaweeds *Cladophora* and *Ceramium* together with gobies, garfish and sand eel larvae (Table 7). In the second time 77 larvae of three gobies species were caught in course of 5-minute haul. In early August after the strong westerly storm this community was destroyed. Only three individuals for three hauls were caught: two larvae of sand goby, one sprat larvae as well flounder fry. Shannon's indices were high but in last case it was only an illusion of the normal state due to likely elimination of sprat larvae. Therefore this was urgent to include only coastal species into the estimation of Shannon's indices in that area.

The comparison of contemporary data with available information from literary sources (Grauman, Lisheva, 1990) demonstrated a noticeable difference between some variables (Table 8). In early 1980s in the Lithuanian coastal zone, attached to northern part of Kaliningrad region, the larval species composition was similar with that from the Russian zone. However the seasonal dynamics of Shannon indices differed. The highest diversity indices were observed in June and July simultaneously with the highest equitability indices and low dominance indices. Almost twenty years later the Shannon's index was the least in July. It was determined the dominance of the only species currently while in 1980s two species have approximately equal abundance (sand goby and lesser sand eel). Grauman and Lisheva reported only the percentage and maximal quantitative values. The highest abundance of sand goby amounted to 20 sp./m^2 in early 1980s, while the maximum value was 160 sp./m^2 in July 2000.

The total species richness in 1992-2002 was higher because the ichthyoplankton composition included apart from of sand goby still three gobies species and also larvae of turbot, garfish, pipe fishes which ones did not occurred in early 1980s. In our sampling two larvae of non-indigenous round goby (length 7-8 mm) were caught in July 2004 and 2015 near Sambia peninsula at depth 9-10 m at the night time probably due to the diel vertical migration inherent in that species at that length range (Hensler, Jude, 2007).

Table 7. Larval species composition, mean abundance (sp./5-min. haul) and Shannon's index Hs at depth 1.0-1.5 m

Date	Weather type	Species composition and abundance							H_s
		Garfish	Sand eel	Sand goby	Common goby	Black goby	Sprat	Flounder	
05-11.07 1999	Long calm	0.6	0.2	5.3	3.5	-	-	-	1.36
24.07. 2014	Long calm	-	-	35.0	22.0	20.0	-	-	1.56
09.08. 2015	After storm	-	-	0.7	-	-	0.3	0.3	1.50

Table 8. The seasonal dynamics of species richness S, Shannon's diversity index Hs, Simpson's dominance index C, Pielow's equitability index E in coastal zone during1980s and 1992-2007

Years	Parameters	Months			
		IV-V	VI	VII	VIII
1980s	S	3	4	4	3
	Dominant species	Herring	Sand goby, sand eel	Sand eel, sand goby	Sand goby
	H_s	1.095	1.611	1.660	1.204
	C	0.571	0.368	0.358	0.460
	E	0.691	0.806	0.833	0.760
1992 - 2007	S	5	4	7	4
	Dominant species	Herring	Sand goby	Sand goby	Sand goby
	H_s	1.544	1.270	0.705	0.844
	C	0.451	0.516	0.742	0.692
	E	0.665	0.635	0.251	0.422

Table 9. Seasonal occurrence and maximal abundance fish larvae in coastal zone in 1946-1955, 1980s and 1992-2007

Species	Years	Seasonal occurrence	Maximal abundance	
			Sp./m^2	Sp./5-min. haul
Snakefish	1946-1955	XII, IV	1	-
	1980-e	-	0	-
	1992-2002	-	0	-
Rock eel	1946-1955	I-IV, VI	7	-
	1980-e	IV	-	-
	1992-2002	III	2	-
Sea snail	1946-1955	II-V	9	-
	1980-e	III-VI	-	-
	1992-2002	III-VII	4	-
Spring-spawning herring	1955	IV-VII	-	600*
	1980-e	IV-VII	20	-
	1992-2002	IV-VII	20	-
Lesser sand eel	1946-1955	V-VIII	33	113**
	1980-e	V-VIII	15	-
	1992-2002	IV-X	6	50
Sand goby	1946-1955	VI-IX, XI	20	-
	1980-e	VI-VIII	20	-
	1992-2002	V-X	160	694

*Grauman, Lisheva, 1990; **Mankowski, 1951.

The literary data on larval occurrence and maximal abundance could be a basis for attempt to reveal long term trends in variability of the coastal larvae assembly (Table 9). In comparison with Polish data for 1947-1955 snakefish larvae disappeared from ichthyoplankton and the duration of occurrence of rock eel declined in the contemporary period. By opinion of Grauman and Lisheva the abundance and number of the sea spawning sites for spring-spawning herring decreased significantly along the SE coast. Among species from the summer spawning origin the abundance of lesser sand eel diminished to the contemporary period. The only species from that list which one sharply raises abundance is sand goby.

Thus, first of all, the tendency is revealed decreasing abundance of larvae of winter- spawning fishes. The larval abundance originated from phytofilous and psammophilous fishes (sand eel and herring) also diminished. The last decline could probably be connected with the drop of stability of their spawning substrata in the sea under the impact of winter - spring storm activity intensification in recent period (Lehmann et al., 2011). Currently the small short - lived sand goby demonstrates the maximal larval abundance and longest seasonal occurrence in the coastal zone (05-60 m) of the SE Baltic. This species is characterized by nest-protection behaviour (Kalinina, 1976). Its spawning sites are often associated with mussel biocenosises (Svetovidov, 1964) on rock grounds and probably less undergone to damage by storm activity. Nowadays the beginning of sand goby reproduction shifted to May from June due to climate warming.

Such the important environmental constraint as eutrophication was described as high for the Gdansk Deep and its coastal zone (Lysiak-Pastuszak et al., 2004). Eutrophication was apparently responsible for changes in macrophytes and macrozoobentos and the structural shift of the ichthyofauna which ones led to domination of Gobiidae in the Polish coastal Puck Bay in the late 1990s (Kruk-Dowgiallo, Szaniawska, 2008). As result at present-day sand goby received a competitive advantage compared with the sea herring and sand eel; the peak abundance of those probably took place in 1950s.

Conclusion

So the mutual trend was observed both in deepwater basins and shallow coastal zone, namely, the transition from polydominant to monodominant ichthyoplankton assemblages. On the open part of the Baltic this was linked to abundance of early developing stages of small pelagic sprat while in the coastal zone with dominance small short-lived bottom species, a namely, sand goby. These changes were similar on the multi-decadal scale (1950-2000s) being determined by climatic and environmental changes. However they probably were in opposite phases in the inter-annual scale. The survival of larval stages of sprat apparently increased due to retention in deep-water spawning sites and formation of frontal zones at those areas under upwelling events in the SE Baltic. On contrary sand goby larval abundance enlarged in conditions of downwelling, while its reproduction suppressed by cold water upwelling inflows. In the coastal zone there was also a tendency to the diminution of occurrence and habitat fragmentation of larval species from winter spawning (snakefish, rock eel) on contrary with dominance of fish larvae originated from summer spawning (gobies).

Shannon's general diversity index estimated by ichthyoplankton data demonstrated decreasing for a few decades both in the open part and the coastal zone of the Baltic. This parameter apparently could indicate some features of ecological status of marine and coastal fish communities in long-term scale.

REFERENCES

Alheit, J., Moelmann, C., Dutz, J., Kornilovs, G., Loewe, P., Mohrholz, V., Wasmund, G. 2005. Synchronous ecological regime shifts in the central Baltic basin and the North Sea in the late 1980s. *ICES J. Mar. Sci.* 62, 1205-1215.

BACC Author Team. 2008. Assessment of climate change for the Baltic Sea Basin. Berlin-Heidelberg, Springer-Verlag, 473p.

Bychkova, I.A., Victorov, S.V. 1987. Elucidation and systematization of upwelling zones in the Baltic Sea. *Oceanology* 27, 2, 218-223.

Elmgren, R., Hill, C. 1997. Ecosystem function at low biodiversity: the Baltic example. In: Marine biodiversity: patterns and processes. Ormond, R. F.G., et al. (eds.). Univ. Press, Cambridge, 319-336.

Golenko, M.N., Golenko, N.N. 2012. Structures of the dynamics fields in the Souheast Baltic Sea during upvelling and downwelling. *Oceanology* 52, 5, 654-667.

Fedorov, V.D., Gilmanov, T.G. 1980. Ecology. Moscow, Mosc. University, 464.

Florin, A.B., Hoeglund, J. 2008. Population structure of flounder (Platichthys flesus) in the Baltic Sea: differences among demersal and pelagic spawners. *Heredity* 101, 27-38.

Henser, S.R., Jude, D.J. 2007. Diel vertical migration of round goby larvae in the Great Lakes. *J. Great Lakes Res.* 33, 295-302.

Hinhrichsen, H.-H., St. John, M., Aro, E., Groenkjaer, P., Voss, R. 2001. Testing the larval drift hypothesis in the Baltic Sea: retention versus dispersion caused by wind-driven circulation. 2001. *ICES J. Mar. Sci.* 58, 973-974.

Grauman, G.B. 1980. Ecological peculiarities of main pelagophillous fishes reproduction in the Baltic Sea. *Fischerei-Forschung* 18, 2, 77-81.

Grauman, G.B, Lisheva, K.M. 1990. Fish spawning in the Lithuanian coastal zone of the Baltic Sea. *Fischerei-Forschung* 28, 2, 39-43.

Kalinina, E.M. 1976. The reproduction and development of gobiids of Asov and Black Sea. Kiev, Naukova Dumka, 120 (in Russian).

Karaseva, E.M., Zezera, A.S., Ivanovich, V.M. 2012. Changes in the species composition and ichthyoplankton abundance along transects in the Baltic Sea. *Oceanology* 52, 4, 478-487.

Karasiova, E.M., Voss R. 2004. Long-term variability of cod and sprat eggs abundance in ichthyoplankton of the Baltic Sea. *ICES CM/*L: 07, 28.

Karasiova, E.M. 2006. Current changes of key environment factors contribution into the Baltic sprat reproductive success as compared to the previous decades. *ICES CM/*C:34. Poster.

Karasiova, E.M., Ivanovich, V.M. 2007. Seasonal and year-to-year variability of ichthyoplankton assemblage diversity indices in coastal zone of the southeastern Baltic Sea under the influence of spatial-temporal environmental dynamics. *ICES CM 2007*/E:07, 20.

Karasiova, E.M., Voss R., Eero, M. 2008. Long-term dynamics in eastern Baltic cod spawning time: from small scale reversible changes to a recent drastic shift. *ICES CM/*J:03, 20.

Koester, F.W., Moellmann, C., Neuenfeldt, S., Vinther, M., St.John, M.A., Tomkievich, J., Voss, R., Hinrichsen, H.-H., Kraus. G., Schnack, D. 2003. Fish stock development in the Central Baltic Sea (1976-2000) in relation to variability in the physical environment. *ICES Marine Science Symposia* 219, 294-306.

Lehmann, A., Getzlaff, K., Harlass, J. 2011. Detailed assessment of climate variability in the Baltic Sea area for period 1958 to 2009. *Climate Research* 46, 185-196.

Lysiak-Pastuszak, E., Drgas, N., Piatkowska, Z. 2004. Eutrophication on the Polish coastal zone: the past, present status and future scenarios. *Mar.Pollut. Bull*, 49, 186-195.

Lindblom, R. 1973. Abundance and horizontal distribution of pelagic fish eggs and larvae in the Baltic Sea 1967-1971. *Meddel. fran Havsfiske-laboratoriet* 240, Lysekil, 38.

Kruk-Dowgiallo, L., Szaniawska A. 2008. Gulf of Gdansk and Puck Bay in the Polish coastal zone. *Egological Studies* 197, Schiever, U. (ed.) Springer-Verlag Berlin-Heidelberg, 139-165.

Mankowski, W., 1950. Macroplanlankton of the Gulf of Gdansk in 1947. *Biuletyn Morsk. Inst. Ryb.* 5, 45-63.

Mankowski, W., 1951. Macroplanlankton of the Southern Baltic in 1949. *Reports of the Sea Fisheries Institute in Gdynia* 6, 83-94.

Mankowski, W. 1955. Macroplankton investigations in Southern Baltic in 1951. *Reports of the Sea Fisheries Institute in Gdynia 8,* 197-233.

Mankowski, W. 1959. Plankton investigations in Southern Baltic in period 1952-1955. *Reports of the Sea Fisheries Institute in Gdynia* 10/A, 69-129.

Moellmann, C., Mueller-Karulis, B., Diekmann, R., Finkman, J., Kornilovs, G., Plikshs, M.,

Axe, P. 2009. The reorganization of a large marine ecosystem due to atmospheric and anthropogenic pressure – a discontinuous regime shift in the central Baltic Sea. *Global change Biol.* 15, 1377-1393

Oeberst, R., Krenkel K., Hinrichs, R., Grauman, G.B. 1981. Vergleich der fangigkeit der ichthyoplankton netze IKS-80, Hensensches eirnetz und WP-2. *Fisherei –Forschung* 19, 2, 53-57.

Odum, E.P. 1971. Fundamental of ecology. Philadelphia-London- Toronto, 740.

Paavola, M., Olenin, S., Leppäkoski, E. 2005. Are invasive species most successful in habitats of low native species richness across European brackish water seas? *Estuarine, Coastal and Shelf Science* 64, 738-750.

Svetovidov, A.N. 1964. Fishes of the Black Sea. Moscow, Nauka, 552.

Soskin, I.M. 1963. Long-term changes of hydrological parameters of the Baltic Sea. Leningrad. Gidrometeoizdat, 159.

Zezera, A.S. 2002. Long-term changes in hydrological characteristics of deep-sea waters of the Southeastern Baltic. *Com.-Biol. Studies of AtlantNIRO in the Baltic Sea* 2, 7-12

In: Baltic and Black Sea
Editor: Andre Mills
ISBN: 978-1-63485-233-3

Chapter 3

ABNORMALITIES IN THE AMPHIDS OF FREE-LIVING NEMATODES IN THE BLACK SEA: IS IT ATAVISM OR RESULT OF A MODERN MORPHO-FUNCTIONAL ADAPTATION TO ENVIRONMENTAL CHANGES AND THE NASCENT PROCESS OF MICROEVOLUTION?

***Nelli G. Sergeeva*[1,*], *Derya Ürkmez*[2,†] *and Tatiana N. Revkova*[3,‡]**
[1,3]Institute of Marine Biological Research of RAS, Sevastopol, Russia
[2]Scientific and Technological Research and Application Center, Sinop University, Turkey

ABSTRACT

In this chapter we aim to summarize our long-term data (1988-2015) about the abnormalities detected in the development of the vital chemoreceptor organs (amphids) of 49 nematode species from the Black

* nserg05@mail.ru.
† deryaurkmez@gmail.com.
‡ alinka8314@gmail.com.

Sea. The total list of species includes 28 valid species and 21 morphotypes, systematic position of which have been identified only to genus level as well as individuals close to a valid species (cf.), however, they require further study and clarification in our view. Amphid abnormalities are characterized by a wide spectrum of appearances. Accordingly, they may have a non-standard number of elements of a species-specific form (instead of two elements they may have 1 or 3-5); one element of amphid may be underdeveloped or not have a species-specific form; the topography of amphids in relation to the front edge of the head and to the axis symmetry of the body may be disturbed. Anthropogenic impact on the Black Sea ecosystem is steadily increasing. Its negative effect is revealed in the extinction of species, community perturbations, serious physiological changes in organisms, resulting in different pathological phenomena. It is obvious that the occurrence of specific abnormalities in the vital organs of the same nematodes species is a result of chromosomal mutagenesis. Repeatability and the wide distribution of species with these types of abnormalities in different regions of the basin are shown in this chapter.

The issues indicated in this chapter require future research: 1. what is the specific cause of the amphid abnormalities in nematodes? 2. Are these amphid abnormalities offer an advantage or a disadvantage compared to the "normal" organs for organisms inhabiting polluted ecosystems or a life under the impact of climate change? 3. Why similar morphological abnormalities are not recorded in nematodes from the other seas? Are these chemoreceptor organ abnormalities in nematodes unique for the Black Sea? 4. Is this phenomenon shows the nascent process of microevolution of nematode fauna in the Black Sea?

Keywords: marine nematodes, morphology, chemoreceptor organ, anomalies, Black Sea

INTRODUCTION

The Black Sea is a deep marginal meromictic sea. It has an oxygenated surface water layer overlying an anoxic deep water layer. The current regime of the Black Sea induces a dome-like shape of the upper layer, with a thickness of the oxygenated surface layer of about 70-120 m in the open sea and of about 150-170 m near the shelf and continental slope [1]. Below, the water (up to 2200 m) is always anoxic and sulfidic. The hypoxic layer of the Black Sea is situated in the lower part of the oxycline, near the hydrogen sulfide zone. In this layer the oxygen deficiency increases with depth.

In addition to the permanent hypoxia found all over the Black Sea, local and seasonal hypoxic zones are also present in shallower areas of the shelf. They are usually associated with increased organic matter input due to natural and anthropogenic eutrophication, whereas some of them are obviously also connected to local methane seepage.

Coastal bottom hypoxic areas are common in different regions of the Black Sea. In the north-western part of the sea, including the Sevastopol Bays, they are reported frequently. Hypoxia is commonly observed in summer, with the strong water stratification and the absence of storms. Reproduction occurs in summer in organisms from within different parts of the ecosystem, which often leads to increased accumulation of organic matter on the bottom. These conditions favor the occurrence of hypoxia, which occurs frequently in the stagnant bottom lows. Coastal bottom hypoxic areas are common in different regions of the Black Sea and in the bays of Sevastopol. Hypoxia is commonly observed in summer, with the strong water stratification and the absence of storms. Reproduction occurs in summer in organisms from within different parts of the ecosystem, which often leads to increased accumulation of organic matter on the bottom. These conditions favor the occurrence of hypoxia, which occurs frequently in the stagnant bottom lows.

The factor contributing to the appearance of the bottom spots of hypoxia were methane seeps from the bottom. By taking into account the peculiarity of the environmental situation, which is created around the methane seeps, we identify hypoxia associated with seepage, as a separate type. Active methane seeps have been discovered in the Black Sea at different depths in recent years.

Phylum Nematoda represents a diverse array of taxa which is present in marine benthic habitats. Marine free-living nematodes live on all substrates with fairly large number of species (the thalli of algae, soft and hard substrates etc.) and at different depths, from the littoral to the abyssal zone. Today, more than 4,000 described species of marine nematodes belonging approximately to 450 genera, 40 families and seven orders [2]. They play an important ecological role in the aquatic ecosystem reaching huge population numbers and species diversity. Due to the diversity of the living conditions and the formative constraints to which these organisms are exposed in the bottom sediments they show pertinent characteristic features to survive successfully in aquatic environments. In the process of evolution, they might have acquired specific metabolism and morphological adaptations that allow them to survive in the bottom sediments under various environmental conditions.

Morphological adaptations, including those related to the structure of the cuticle, the form of head, number of somatic and cephalic seta, special lateral tactile organs (amphids), the structure of the stoma, digestive organs, form and length of the tail and several others are specific characteristics for each species. Nematodes have the ability to show trophic adaptations to diverse derivative habitats. Free-living nematodes can withstand stresses of various origins by undergoing cryobiosis, osmobiosis, anoxybiosis as well as thiobiosis using the detoxification mechanism of sulfide.

In general, the nematodes are morphologically very variable. Individuals of one species from different populations can vary significantly among themselves and the most variable character is the absolute body size.

Teratological published data about free-living nematodes from other seas are not numerous however they show various possible deviations from the standards. Most of the recorded anomalies involve the reproductive system (gynandromorphism, hermaphroditism, increased or decreased number of reproductive tubes, bivulvularity, etc.) [3–9].

Abnormal development of double posterial oesophageal bulb in nematodes was first discovered in *Theristus* sp. from the coastal zone of Sinop Bay (Turkey, Black Sea) (Ürkmez and Sergeeva, unpublished data).

V.V. Galtsova [10] studied the structure of amphids in the species of *Sabatieria*, and found both the ventral and dorsal spiral twistings. The author concluded that enantiomorphism of spiral amphids of nematodes is quite a variable sign. G.G. Gagarin [11, 12] considers that the localization of nematode amphids is so uncertain and he doubts the validity of this character for species identifications.

Nematodes with four amphids were reported in two papers. Four large chitinizated rings were described on the head terminal of a specimen from the genus *Pseudochromadora*. The lateral two were taken from for amphids and the two medial ones from something else [13]. Two pairs of full-grown spiral amphids, one beneath the other was described for *Achromadora terricola* (De Man 1880) [14]. All these anomalies involved only single individuals.

Scientists attributed several functions to the amphids of nematodes. Some researchers consider them as a primary chemoreceptor organ [15, 16, 17]. Others believe that amphids also have mechano - and photoreceptor functions [18, 19]. V.V. Galtsova [10] accepts the multifunctionality of amphids, with one main and several secondary functions.

The most conservative features of nematodes are the structure and the form of amphids (lateral organs). In the evolution process, the marine and freshwater free-living nematodes have acquired two specific elements of chemoreceptor organs that are peculiar to each individual species, regardless of geographic and habitat distribution. Form of amphids and their orientation relative to the symmetry axis and the front edge of the head is also specific for each species. Amphid is located usually on the front part of the head and has a pocket-type hollow immersed in the cuticle, serving as an "olfactory cavity which consists a–powerful beam of chemoreceptor nerve fibrils inside the pocket." The position, size and symmetry of the (fovea) amphids are specific characters not only for each species, but also for the genus, and used in the systematics of free-living nematodes.

Amphids are composed of two identical elements in the form of spiral, round, can be slit-like, in the form of a pocket, in the form of a loop or in the form of a horseshoe, etc. They are especially strongly developed in males, that in often pronounced as part of sexual dimorphism in free-living nematodes. V. Galtsova [20] considers a wider range of functions of amphids in nematodes, namely along with the chemoreception, they are suggested to play the role of organ allocation.

In this chapter, we aim to summarize our long-term data (1988-2015) about the abnormalities detected in the development of the vital chemoreceptor organs (amphids) of nematode species found in the Black Sea. We hope that further interdisciplinary studies which are described anomaly in the development of nematodes as a phenomenon for the Black Sea will be found in other water basins that will lead to understanding their origin and their role in the evolutionary processes.

MATERIAL AND METHODS

We include protozoa and multicellular animals in meiobenthos, as it is considered by the researchers of meiobenthic communities [20–23]. Our investigations cover meiobenthos and nematode fauna at different zones of the Black Sea from the coastal shallow waters to the permanently anoxic and hydrogen sulfide containing deep-waters. Materials and methods of collecting and analyzing meiobenthos are described in a number of our publications [23–26].

The taxonomic nematodes classification of WoRMS online database (http://www.marinespecies.org, Accessed 2016-02-20) are used [27].

New information about the spatial distribution of free-living nematodes with clear chemoreceptor organ abnormalities was obtained in the framework of several projects such as the "Crimea" project, FP 6 EU "Hermes," FP 7 EU "Hypox," "Perseus" and "CoCoNet" projects, TUBITAK-NASU project and National Programs of NAS Ukraine and Russian Academy of Sciences. Deep-sea meiobenthos samples collected from the North-Western part of the Black Sea in 2003, 2004, 2007 and 2010, and from the South-Western (Istanbul Strait's outer area) part of the Black Sea in 2009 and 2010 were studied. Nematodes from the *Phyllophora* fields in Zernov and the Karkynitsky Bay (NW, Crimean shelf) and area near the Kerch Strait (SE, the shelf of the Crimea) were studied in bottom sediment samples which were obtained during 2010-2011 and 2013 sampling seasons, respectively. Free-living nematodes obtained from the coastal waters of the Sevastopol region (Sevastopolskaya Bay including Yuzhnaya bay and the mouth of Chornaya river (near Inkerman) in Sevastopol, Kruglaya and Balaklavskaya Bays) and the hyperhaline Sivash Lake (north of the Crimean Peninsula) were studied in detail during the sampling seasons 2005–2013, 2013 and 2015, respectively (Figure 1).

Figure 1. The study areas in the Black Sea where abnormality of free-living nematodes were found: I – Varna Bay (1988, 1989), II – Danube estuary (1990, 1994), III – Donuzlav Lake (1999), IV – Oxic/Anoxic interface along the methane seeps of the Dnieper paleo-delta Canyon (2003, 2004 and 2007), V – Oxic/Anoxic interface in shelf and upper slope western Crimean peninsula (2010), VI – Karkynitsky Gulf and *Phyllophora* Field of Zernov (2010-2011), VII – North-eastern part of the Black Sea area near the Kerch Strait (2013), VIII – Sevastopol Bays (2005 –2013), IX – Sivash Lake (2013, 2015); X– Shelf and upper slope of Turkish areas (1999, 2009 and 2010, 2013).

A large field of methane seeps, the submarine Dnieper Canyon is located on the northwestern part of the Black Sea [28, 29]. Samples were collected from ten stations at this area crossing the shelf and the NW Crimean Peninsula slope characterized by methane seeps along an oxic/anoxic interface between water depths of 120 – 240 m during the M 72/2 MICROHAB cruise with the German RV Meteor (February-March 2007). The undisturbed sediment samples were obtained using a multi-corer.

Earlier in 2003 and 2004, sediment samples from three stations at depths of 94, 98 and 188 m obtained during the RV 'Poseidon' (Germany) cruise and RV 45, 58 and 60 'Professor Vodyanitsky' (Ukraine) cruises were transferred to our colleagues from IBSS to study meiobenthos from this area.

Samples were collected from 21 stations at water depths ranging between 80-375m in the north-western part of the Black Sea along a transect of an oxic/anoxic interface (near Crimean Peninsula) in April–May 2010 during the 15/1 Hypox cruise with the German RV 'Maria S. Merian'. Meiofaunal organisms were studied in the upper 5 cm sediment horizons of two–four cores per station, with each core covering an area of 70.9cm^2 (TVMUC) and 41.8cm^2 (JAGO pushcore).

Bottom sediment samples from the *Phyllophora* fields in the Zernov and Karkynitsky Gulfs, in the north-western part of the Black Sea were collected in the same way during RV "Professor Vodyanitsky" cruises 70 and 72 in November 2010 and August 2011. Meiobenthos was studied at 27 stations from 10 to 50 m water depths. Samples were taken by bottom grab "Ocean 50" using the meiobenthic sampling tubes with 3 columns covering a 18.1 m^2 of bottom area. Sampling material was analyzed in the laboratory of the Department of Benthos Ecology in IBSS.

Near the Kerch Strait region, in the north-eastern part of the Black Sea, bottom sediment samples were also collected during RV "Professor Vodyanitsky" cruise 75 in July 2013. Bottom sediments collected from 16 stations in the range of a 60–249 m water depth using a gravity-corer with a diameter of 127 mm.

Sediment samples from Sevastopol bays were collected in 2005, 2009, 2010 and 2013 by Scuba Divers or a "Petersen's" bottom grab and using a meiobenthic pushcore (18.1 cm^2 cross-sectional area) at 5 m depth. The material was collected in five bays in the vicinity of Sevastopol, which differ in the nature and degree of anthropogenic pressure. All samples were collected as two or three replicates.

In the southwestern part of the Black Sea (Istanbul Strait, Turkey) samples were obtained from 9 stations with a depth range of 75–300m during the

Hypox cruises with RV 'Arar' (ITU) in November 2009 and from 13 stations during the cruise of 15/1'Maria S. Merian'(German) in April 2010. Meiofaunal organisms were studied in the upper 5 cm of the sediment samples taken as two–four cores per station, with each core covering an area of 70.9cm^2 (TV multi-corer and Gravity-corer).

Meiobenthic samples were also collected monthly at 8 stations (3–10 m) located at the central part of the Southern Black Sea (Sinop, Turkey) in the scope of a joint project between TÜBİTAK and NASU conducted between August 2009 and July 2010 [26]. Samples were taken as three replicates by Scuba Divers using metal push cores with a mouth area of 12.56 cm^2.

All samples were preserved in 75% alcohol. As we know from our previous experiences, this percentage of alcohol preserves morphological structures without distortion. We avoided prior fixation in formalin in order not to damage calcareous taxa. The sediment was carefully washed on sieves with mesh sizes of 1 mm and 63 μm in the laboratory. The fractions retained on the sieves were stained with Rose Bengal solution before being sorted in water under a binocular microscope for "live" (stained) organisms, which were then identified to higher taxa. Intensely stained organisms not showing a sign of morphological damage were extracted from the 1.000 to 63 μm size fractions using a Bogorov counting chamber. All the isolated organisms were counted and identified to higher taxa or to species level when possible. The specimens were placed in cavity slides with a mixture of glycerol (50%) and water (50%). From each sample did a random sampling for 150 to 200 specimens of nematodes for analysis of species composition of community. During 1998-2015, morpho-anatomical analysis of all specimens was carried out with a magnification of x1000-1300 using microscopes MBS-3, MBS-6, Canon and Olympus CX41. Photographs were taken using a Canon A-620 digital camera and an Olympus E-410 camera attached to Olympus CX41 compound microscope.

PREVIOUS RESEARCH

Earlier in the course of long-term investigations of the Black Sea meiobenthos, including nematodes, morphological deviations were no observed. In 1988, the first described population of *Terschellingia longicaudata* was collected at 77 m depth in aleurite ooze with an admixture of *Modiolula phaseolina* (Bivalvia) shells of the Bulgarian shelf. The upper oxidized layer of the bottom sediment (0.5-0.8 mm) was brown with a thick sour-cream like structure. Below it (down to 3cm), a layer of grayish-brown ooze occurred, more dense, but plastic. Beneath 3-5 cm, the dark grey, black spotted ooze (reduced-sediments) was replaced by the dense black ooze with obvious strong hydrogen sulfide smell.

The nematode fauna consisted of 42 species and *T. longicaudata* was among the dominant species. The amphids of the genus *Terschellingia* are usually arranged along the lateral line and look like a ring with more or less thick walls. The peculiarity of the arrangement, size and the fovea symmetry of the amphids are important specific signs. The abnormal individuals made up 33.3% of the total population. The abnormalities are reflected in the number, arrangement and localization of the chemoreceptor organ ('lateral bodies' or 'amphids') and they revealed differences from the normal. These amphid disorders were registered in adults (male and female) and juveniles. The variations of abnormal amphids were studied according to the morphological variability of principal characteristics of the species in the natural populations [30, 31].

Number of amphids in abnormal specimens varied from 3 to 5. The studied population of *T. longicaudata* collected from the Bulgarian coast included specimens with number of amphids ranging from 2 (norm) to 5. Diameters of all the rest ("accessory") were either near to standard or smaller. The "accessory" amphids were removed 4.0–50.0 μm from the anterior head margin. The amphid orientation varied within the head end limits. They were arranged laterally, sublaterally, dorsally, ventrally in the same specimen. One pair of amphids had the size and localization typical for the species (Figure 2).

Since then, we constantly observe anomalous nematodes in the Black Sea. This made it possible to install the analogical type of abnormal development in nematodes from different coastal areas of the Black Sea under chronic human impacts, and in areas with methane seeps on the sea floor.

The studies of the changeability of signs continued in 1990–1995 [30–32] when we investigated meiobenthos, inhabiting coastal zones with high anthropogenic impact (Danube estuary, the Donuzlav Bay, Sevastopol bays, three ports Rize, Trabzon, Ordu-Fatsa of Turkey). As a result, the list of nematode species with disturbances in amphid development reached to 10 more species (*T. longicaudata, Monhystera* sp., *Metalinhomoeus zosterae, Eleutherolaimus longus, Sabatieria pulchra, Halichoanolaimus dolichurus, Axonolaimus setosus, Microlaimus* sp., *Parodontophora quadristicha* and *Paracanthonchus caecus*).

In contrary to the Bulgarian shelf, abnormal specimens were found in the Donuzlav Bay in six nematode species. Disturbances in the development of amphids were mainly represented by phenotypes, described for the Bulgarian shelf. Anomalies were marked in *A. setosus* and *P. quadristicha* and should be underlined. One specimen of *A. setosus* had one amphid of a species-specific form, and another was in the form of two ovals. In the male of *P. quadristicha* one amphid was also species-specific, while another had a form, non-peculiar for this genus. Accumulated data relating to abnormal development of nematodes in different water bodies of the Black Sea were used to develop a method in determining the toxicity of marine sediments.

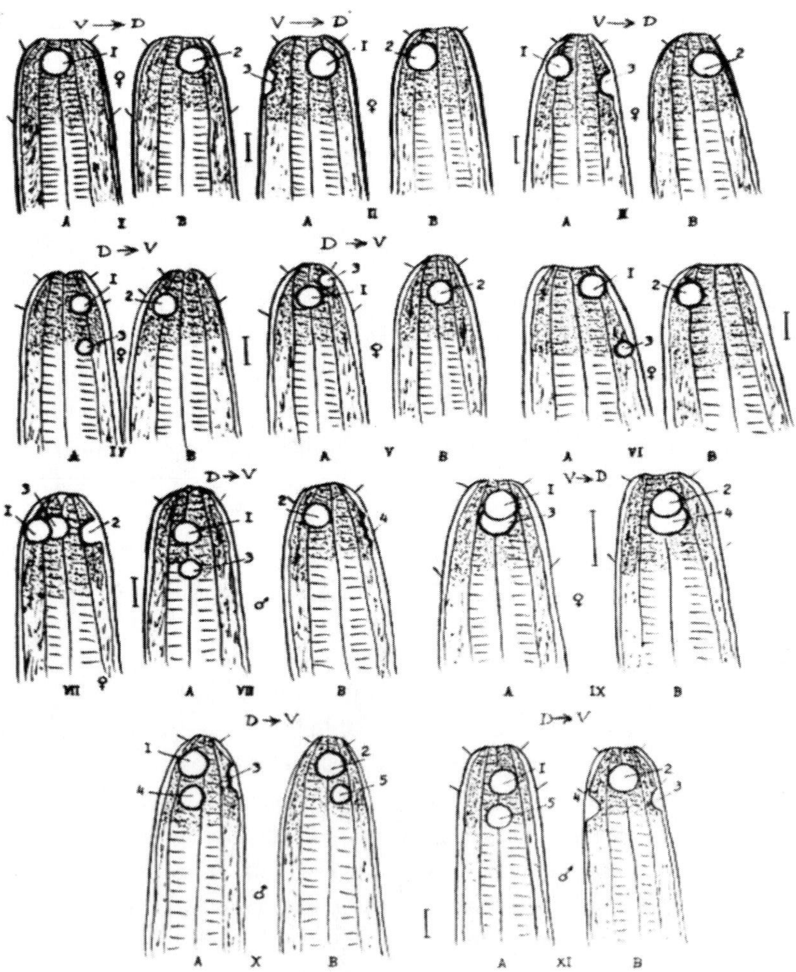

Figure 2. Variations in number and topography of the amphids in *Terschellingia longicaudata* found in 1988 in the Varna Bay [by 30, 31].

Share of abnormal specimens in total nematode populations reach to 1.3–50% And their density at a square unit extended to considerable values. For example, the abundance of *T. longicaudata* with morphological deviations is 1.2–2.0 thousand ind.*m^{-2} and the aberrant specimens of *A. setosus* make 4.0 thousand ind.*m^{-2}. Thus, registration of abnormal specimens in the Donuzlav Bay can not be considered as occasional. Individuals with abnormalities made 11.2 to 12.7% of the total species populations in three Turkish coastal areas in the Black Sea (Rize, Trabzon, Ordu-Fatsa) [32].

As in other regions of the Black Sea, abnormal specimens were also found in the Danube region. These abnormal specimens make 3.2–6.3% of the total population of *S. pulchra*, 6.3% of the total population of *A. setosus*, 1.6–7.5% of the total population of *T. longicaudata*, and 25,0% of the total population of *Monhystera* sp. The abnormal specimens of these species comprised 1–13% of total populations in the Balaklava and Inkerman Bays in the Sevastopol region (Table 1). All above mentioned disturbances in the development of amphid early were found only in the regions with anthropogenic (technogenic) impact and in methane seeps area, so they have been used as indicators for the assessment of the quality of sediments [30, 32].

Previously, 11 phenotypes of a single species of nematode (*Terschellingia longicaudata)* have been described on the basis of the variability in amphid development [30, 31]. Further study of this phenomenon led us to the conclusion that amphid abnormalities s in the Black Sea are characterized by a wide spectrum: 1) they may have non-standard number of elements of a species-specific form (they may have 1 or 3–5 elements instead of two); 2) specimens may have two amphids, but with different forms; 3) one element of amphid can be underdeveloped or may have not a species-specific form; 4) four amphids composed of double species-specific elements; 5) the topography of amphids in relation to the front edge of the head and to the axis symmetry of the body may have been disturbed [32].

In this report, our goal is to generalize our existing previously obtained data and provide new information on the abnormalities seen in the structure and composition of the vital organ (amphid) and its unusual topography in body of many nematode species as well as to show spatial distribution of this surprising phenomenon in the Black Sea.

NEW RESULTS AND DISCUSSION

We provide here the summary results of a long study period (1988–2015) on free-living nematode fauna of the Black sea which brought unexpected information about the existence of unusual abnormalities in the development, layout and localization of the amphids in the body of nematodes. To date, our collections from different regions of the Black Sea are represented with individuals of 49 nematode species with various teratisms in structure of amphids. The total list of species (Table 1) includes 26 valid species and morphotypes, systematic position of which have been identified only to genus level as well as individuals close to a valid species (cf.), however, they require further study and clarification in our view. It should also be emphasized that, this phenomenon is widespread in the Black Sea at present time, and is not limited to the habitats with high technogenic pressure or to certain environments such as methane seeps, hydrogen sulfide zones, hypoxic bottoms, sediments contaminated with petroleum hydrocarbons and heavy metals, hyper saline waters, etc).

Table 1. The total list of free-living nematode species with amphid abnormalities from the studied areas of the Black Sea

Species	Study area									
	I	II	III	IV	V	VI	VII	VIII	IX	X
Chromadorida Chitwood, 1933										
Halichoanolaimus dolichurus Saweljev,1912		+	+	+						
Paracanthonchus caecus (Bastian, 1865)				+						+
Araeolaimida Sch.-Stekh. et Coninck., 1933										
Araeolaimus sp.1				+						
Axonolaimus ponticus Filipjev, 1918			+					+		
A. setosus Filipjev, 1918			+	+	+			+		
Campylaimus sp. 4				+						
Campylaimus sp. A				+						
Diplopeltis onustus Wieser, 1956						+				
Paracomesoma dubium (Filipjev, 1918)								+		

Species	Study area									
	I	II	III	IV	V	VI	VII	VIII	IX	X
Parodontophora quadristicha (Sch.- Stekh. 1950)			+		+			+		+
Pseudolella sp.1								+		
Sabatieria sp. A						+				+
Sabatieria sp										+
Sabatieria abyssalis (Filipjev, 1918)				+	+	+	+	+		+
S. longicaudata de Man 1907				+						
S. praebosporica Sergeeva, 1973										+
S. pulchra (Schneider, 1906)			+	+	+	+	+	+	+	+
Desmodorida De Coninck, 1965										
Aponema sp.								+		
Desmodora conica Vitiello, 1970			+	+						
D. pontica Filipjev, 1922				+						+
Microlaimus kaurii Wieser, 1954								+		
Microlaimus sp. M				+						
Microlaimus sp. P				+						
Microlaimus sp.1			+		+					
Microlaimus sp										+
Pseudomicrolaimus murinae, Sergeeva 1976				+						
Spirinia parasitifera (Bastian,1865) Gerlach, 1963								+		
Molgolaimus sp.								+		
Monhysterida Filipjev, 1929										
Eleutherolaimus longus Filipjev, 1922								+		
Linhomoeus sp. A				+						
Metalinhomoeus sp. O				+						
M. zosterae Filipjev, 1918			+		+	+	+			
Monhystera sp. A								+		
Paralinhomoeus sp.				+						
P. deconincki Groza-Rojancovsky, 1972						+	+			
Sphaerocephalum crassicaudata Filipjev, 1918								+		
Sphaerolaimus ostreae Filipjev, 1918								+		

Table 1. (Continued)

Species	Study area									
	I	II	III	IV	V	VI	VII	VIII	IX	X
Sph. gracilis de Man 1876	+			+				+		+
Sph. dispar Filipjev, 1918							+			
Sph. macrocirculus Filipjev, 1918							+			
Terschellingia. sp.P				+						
T. longicaudata Filipjev, 1922	+	+	+	+	+	+	+	+		+
T. cf. *longicaudata* Filipjev, 1922	+						+	+	+	+
Theristus maeoticus Filipjev, 1922								+		
Th. sabulicola Filipjev, 1918								+		
Theristus sp. A				+						
Plectida Gadea, 1973										
Diplopeltoides sp.1							+			
Leptolaimus cf. *steineri* (Filipjev, 1922)						+	+			

*Note: The study areas in the Black Sea where abnormality of free-living nematodes werefound: I – Varna Bay. (1988, 1989), II – Danube estuary (1990, 1994), III – Donuzlav Lake (1999), IV – Oxic/Anoxic interface along the methane seeps region of the Dnieper paleo-delta (2003, 2004 and 2007), V – Oxic/Anoxic interface along shelf and upper slope western Crimean Peninsula (2010), VI – NW, Karkynitsky Bay and *Phyllophora* Field Zernov's (2010-2011), VII – North-eastern part of the Black Sea near the Kerch Strait area (2013), VIII – Sevastopol Bays (2005-2013), IX – Sivash Lake (2013, 2015); X – Shelf and upper slope of the Turkish areas (1995, 2009 and 2010).

Oxic/Anoxic Interface along the Methane Seeps Region of the Dnieper Paleo-Delta

The area of Dnieper paleo-delta is the most prolific seep area of the northwestern Black Sea margin and possibly one of the most active in the world [33]. The studies conducted at the methane seeps in this area resulted in discoveries of more seeps [28, 34] as well as revealed the presence of carbonate build-ups formed by anaerobically oxidized methane in an anoxic environment at 230 m water depth [35, 36, 37 and 38]. In the vicinity of the submarine Dnieper Canyon (northwestern part of the Black Sea), a large area of nearly 2000 km^2 which is found to be covered by hundreds of methane seeps. Also, massive carbonate reefs with thick microbial mats were discovered at methane gas seeps deep in anoxic waters. At present, there is lack of information concerning bottom fauna composition in areas with oozing of methane gas in the Black Sea [39, 40 and 41]. Therefore, studies of benthos to be carried out at the methane jet oozing areas are one of the actual tasks of marine ecology.

On the other hand, recent detections of benthic fauna in the Dnieper Canyon raise question about oxygen supply, and limitations for benthos oxygen demand [23, 42]. These observations make the Crimean shelf a highly suited area to study oxygen availability and the environment response over extended periods of time.

The study of deep-sea meiobenthos conducted in 2007 near the Dnieper underwater Canyon (NW, Black Sea) along transect of 120–240 m revealed 16 high level taxa of meiobenthos and free-living nematodes were the dominant taxon [23].

Our preliminary data show that species diversity at depths between 120 and 240 m includes 6 orders and 23 families. Some nematodes can live in permanently anoxic environments due to a high tolerance to hydrogen sulfide conditions. The nematode fauna is unique at the studied depths. It includes 90 species and 9 genera unknown from the Black Sea, and 19 species, 9 genera, and 1 family recognized for the first time in this basin. The fauna includes stenobiont and eurybiont forms adapted to live in the redox zone. The species richness of free-living nematodes decreased with increasing depth, highest values being found at depths of 120 and 130 m with a second peak in number of species and genera located at a depth of 170 m (O_2 = 0.07 mmol L^{-1}, sulfide = <1 μmol L^{-1}). Species adapted to anaerobic conditions occurred at depths ranging from 190 to 240 m [23].

A total of 90 nematode species were found. Among the populations of 12 nematode species, 10 specimens collected from 7 stations were found to have distinct morphological abnormalities in their chemoreceptor's organs. Structural and topographical abnormalities were recorded for the amphids of mature female and male individuals.

Table 2 shows the presence of abnormal mature specimens in the populations of 12 species and morphotypes in the depth range of 120–180 m. We frequently noted amphid abnormalities in this area. Individuals were observed to have one, three or four elements of amphids, instead of their standard number as two.

Table 2. Share of abnormal and normal individuals in the populations of nematode species found at the column of seepage area bottom sediments at the Oxic/Anoxic interface in the methane seeps region of the Dnieper paleo-delta (RV '*Meteor*', 2007)

Species	Number of amphids	Station/Depth, m															
		RV 'Prof. Vodyanitsky' 2004		RV 'Meteor, 2007'													
		St. Dnieper 1/98		256/120		258/130		259/140		260/150		261/160		262/170		324/180	
		ind.	%	ind.	%	ind.	%	ind.	%	ind.	%	ind.	%	ind.	%	ind.	%
Linhomoeus sp. A	4 2*	-	-	-	-	-	-	-	-	1♀ 20	4.8 95.2	-	-	-	-	-	-
Desmodora conica	1 2*	-	-	-	-	1♀ 13	7.2 92.8	-	-	-	-	-	-	-	-	-	-
Metalinhomoeus sp.	4 2*	-	-	-	-	-	-	1♀ 26	3.8 96.2	1♀ 31	3.2 96.8	-	-	-	-	2♀ 0	100 0
Paralinhomoeus sp.	3 2*	-	-	-	-	-	-	2♀ 29	6.5 93.5	-	-	-	-	-	-	-	-
Sabatieria abyssalis	4 2*	-	-	2♀ 81	2.5 97.5	-	-	-	-	1♀ 61	1,7 98.3	-	-	-	-	-	-
S. pulchra	4 2*	-	-	3♀ 118	2.5 97.5	-	-	-	-	-	-	-	-	-	-	-	-
Campylaimus sp. A	4 2*	-	-	-	-	-	-	-	-	2♀ 3	40.0 60.0	-	-	-	-	-	-
Campylaimus sp. 4	2change 2*	-	-	-	-	-	-	-	-	-	-	2♀ 7	22.3 77.7	-	-	-	-
Microlaimus sp. M	4 3 1 2*	-	-	-	-	-	-	-	-	-	-	1♀ 1♂,1♀ 1♀ 64	1.4 3.0 1.4 94.2	-	-	-	-

Species	Number of amphids	Station/Depth, m															
		RV ‘Prof. Vodyanitsky’ 2004		RV ‘Meteor, 2007’													
		St. Dnieper 1/98		256/120		258/130		259/140		260/150		261/160		262/170		324/180	
		ind.	%	ind.	%	ind.	%	ind.	%	ind.	%	ind.	%	ind.	%	ind.	%
Microlaimus sp. P	1			-	-	-	-	-	-	-	-	2♀	2.2	-	-	-	-
	3											-	-	2♀	7.7		
	2*											89	97.8	24	92.3		
Pseudomicrolaimus murinae	4	1♀,1juv	50.0	-	-	-	-	-	-	-	-	-	-	-	-	-	-
	2*	2	50.0														
Terschellingia sp.P	3	-	-	-	-	-	-	-	-	-	-	1♀	20.0	-	-	-	-
	2*											4	80.0				
Theristus sp. A	4	-	-	-	-	-	-	-	-	-	-	2♀	1.4	-	-	-	-
	2*											145	98.6				

*Note: normal species-specific form and number of amphids.

The greatest variability in the number of amphid elements were found in *Microlaimus* sp. M. Additionally, two females belonging from seven specimens of *Campylaimus* sp. 4 were found to have amphids other than their species-specific form, accounting for 23% of the population. Such abnormal representatives were also found in *S. abyssalis* and *Metalinhomoeus* sp. with the character of the anomaly observed in the specimens collected from different water depths. Figure 3 demonstrates the abnormal development of amphids in two species *Microlaimus* sp. M. and *Campylaimus* sp. 4 from habitats characterized by methane releases from bottom sediments. Despite the fact that the proportions of the individual abnormal specimens on little squares of every studied sample are usually small, it should be noted that their number will be more significant in the total abundance populations at m^2 of the bottom area.

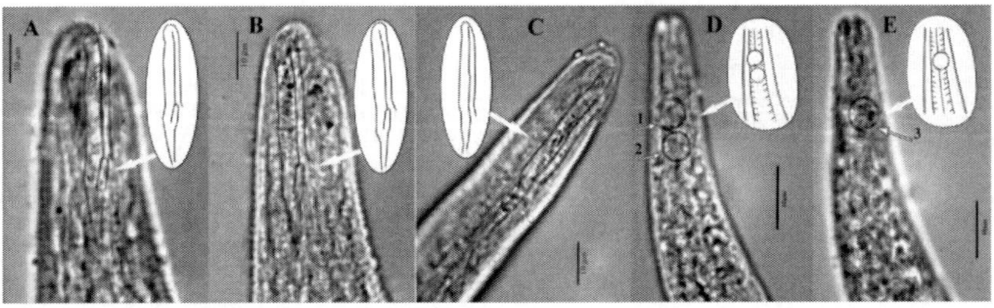

Figure 3. Variety in abnormal development of amphids in some species of nematodes found at the submarine Dnieper Canyon; A–C *Campylaimus* sp.4: A–B One specimen with two modified elements of amphid, C – amphid with species-specific form; D–E *Microlaimus* sp. M: The individual has three species-specific amphids instead of two elements.

Note that, one of the two female specimens of *Paralinhomoeus* sp. with four amphids were found in bottom sediment sample collected at a depth of 188 m (St. 802) during the cruise of RV '*Poseidon*' (Germany) in 2004. Two female specimens of *Pseudomicrolaimus murinae* among a total of four were also observed to have four amphids (RV '*Professor Vodyanitsky*', cruise 60, 2004, project "Crimea," Germany) at a depth of 98m (St. Dnieper 1). Examination of bottom sediment (0-1cm) samples collected from the center of methane seeps at a depth of 94 m (St. 4) revealed one female among two individuals of *Desmodora pontica* and one female among 7 specimens of *T. longicaudata* with four amphids. The below layer (1–2 cm) presented with two specimens of *S. abyssalis* (from a total of 9 individuals) and two specimens of *Sph. gracilis* (from a total of 2 individuals) also with four amphids.

Oxic/Anoxic Interface along the Shelf and Upper Slope of the Western Crimean Peninsula

Hypoxia occurs where oxygen concentrations fall below a physiological threshold of many animals, usually defined as <63 $\mu molL^{-1}$. Oxygen depletion can be caused by anthropogenic influences, such as global warming and eutrophication, but as well occurs naturally due to restricted water exchange in combination with high nutrient loads [43].

Recordings of bottom-water oxygen concentrations along transect from 95 to 218m water depth provided us a differentiatiation of four zones of different bottom-water oxygenations within a distance of more than 30 km. These are the "oxic zone" at water depths of 95 to 130m, "oxic–hypoxic zone" at water depths between 130 and 142m with average oxygen concentrations of 116±29 μmolL, "oxic–hypoxic zone" at water depths between 130 and 142 m, with average bottom-water oxygen concentrations of 94±56 $\mu molL^{-1}$, "hypoxic–anoxic" zone between 142 and 167 m water depth sediments with fluctuating hypoxic conditions between 0 and 63$\mu molL^{-1}$. The bottom water below 167m was permanently anoxic during the study period, sulfide was constantly present in the bottom water below 180 m, with concentrations ranging between 5 and 23 $\mu molL^{-1}$. In this "anoxic–sulfidic" zone, sediments were blackish-dark green [44].

The sediment and the water column were sampled along transect from 95 to 375m water depths. Analysis of meiobenthos and nematode species composition along all depth ranges has not yet been completed. A new study of meiobenthos along transect of 80–163m revealed 18 higher level of meiobenthic taxa, in which free-living nematodes were dominant. Three species and three morphotypes of nematodes with distinct amphid anomalies were recorded at 6 stations at this depth range (Table 3). All the species with amphid anomalies showed four elements of amphids except for *Terschellingia longicaudata* which exhibited a wider range of anomalies. For example, the populations of *T. longicaudata* made up 80% and 19.2% of all abnormal specimens at depths of 104 m (St. 464) and 101 m (St. 482), respectively. The number of amphids with species-specific size and shape ranged between one to four in one case, whereas it consisted of four amphids with different size or with an undeveloped or modified form of one amphid in another case (Table 3).

Table 3. Share of abnormal and normal individuals in the populations of nematode species found at the Oxic/Anoxic interface along the shelf and upper slope of the western Crimean Peninsula (RV '*Maria S. Merian*', 2010)

Species	Number of amphids	Station/Depth, m											
		372/163,2		378/155,4		423/124		424/148		464/104		482/101	
		ind.	%	ind.	%	ind.	%	ind	.%	ind.	%	ind.	%
Cobbia sabulicola	4	1♀	100	-	-	-	-	-	-	-	-	-	-
	2*	0	0	-									
S. abyssalis	4	-	-	1♀	8.3	1♀	5.0	-	-	-	-	-	-
	2*			11	91.7	19	95.0	-	-	-	-	1♀	16.7
	change	-	-	-	-	-	-					5	83.3
Monhystera gen. sp.	4	-	-	1♀	100	-	-	-	-	-	-	-	-
	2*			0									
Metalinhomoeus sp.1	4	-	-	-	-	-	-	1♀	4.8	-	-	-	-
	2*							20	95.2				
T. longicaudata	4	-	-	-	-	-	-	-	-	1♀	20.0	1♂, 2♀	11.6
	4 + differ.size	-	-	-	-	-	-	-	-	-	-	1♂	3.8
	2*	-	-	-	-	-	-	-	-	1	20.0	21	80.8
	3	-	-	-	-	-	-	-	-	1♂	20.0	1♀	3.8
	1	-	-	-	-	-	-	-	-	1♂	20.0	-	-
	undevelop	-	-	-	-	-	-	-	-	1♂	20.0	-	-
Ascolaimus cf. *elongatus*	4	-	-	-	-	-	-	-	-	-	-	1♀	12.5
	2*	-	-	-	-	-	-	-	-	-	-	7	87.5

*Note: normal species-specific form and number of amphids.

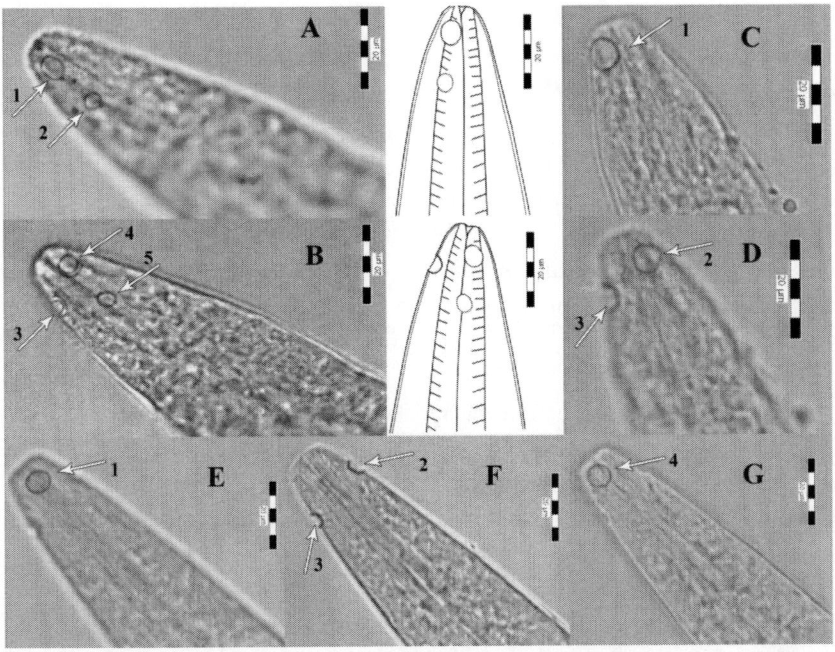

Figure 4. Variety in number and topography of abnormal amphids in *Terschellingia longicaudata* found at the Oxic/Anoxic interface along the shelf and upper slope of the western Crimean Peninsula (RV '*Maria S. Merian*', 2010). A-B one specimen with five species-specific elements of amphid, C-D one specimen with three species-specific elements of amphid, E-G one specimen with four species-specific elements of amphid.

A few specimens with special structure disturbances were also recorded along with the usual amphid anomalies found in these areas. Figure 4 demonstrates the peculiar disturbance in the amphids of one female specimen of *S. abyssalis* and one male specimen of *T. longicaudata.*

Karkynitsky Gulf and the Zernov *Phyllophora* Field

Karkynitsky Gulf is a very important area as a region of high biological and recreational resource. It serves as the main habitat for sturgeon in the north-western part of the Black Sea. In the area of Karkynitsky Gulf, the "Swan Island" Reserve and aquatic complexes, providing recreation services are located. Despite the existence of protected areas, the Gulf is experiencing significant anthropogenic stress in connection with the activities of large industrial and agricultural enterprises on the coastal zone. From the second-half of the 1980s, the climatic decrease in the water content of rivers was prominent. The speed of flows in estuarine regions decreased it is more than 2 times in comparison with earlier years [45]. All these processes strengthened the stratification of waters and weakened the processes of vertical circulation, which contributes to the appearance of hypoxia.

Along the Karkynitsky Gulf, uneven distribution of physical and chemical properties of bottom sediments, with change of oxygen concentrations and nutrients in the environment, salinity, temperature and of some biotic factors, such as the state of the populations of *phyllophora* and availability of macrobenthic fauna are observed.

Increase in the stratification of density and reduction of vertical exchange by oxygen was descended due to the climatic factors. Hydrological regime is one of the leading factors, which determine the state of shelf bottom communities. Conditions of hypoxia, which lead to the anoxia phenomena originated from the specified hydrological conditions.

This field is a unique accumulation of *Phyllophora* red algae (water-plant), located in the center of the northwestern part of the Black Sea. As it is known, notable changes were observed in the benthic ecosystems of the north-western Black Sea as a result of eutrophication and irrational use of natural resources in the second half of the last century.

The enormous volumes of *Phyllophora* present in the first half of the XX century diminished sharply in 80–90 years. The abundance and biomass of *Phyllophora* reduced in 100–150 times. Degradation of biocenosis of *Phyllophora* in Zernov takes place due to the influence of river runoff [46].

In 1991, meiobenthos of the soft bottom sediments of Zernov *Phyllophora* field was found to be composed of 13 higher taxa (type, class, order) and its density was a little higher than 760 thousand ind.•m^{-2}. A complex of Nematoda + Foraminifera dominated in the mean density of meiobenthos of this area [47].

Modern taxonomic structure of meiobenthos was studied in 2010–2011 during 68 and 70 cruises of RV "Professor Vodyanitsky." Bottom sediments were collected at 26 stations from 10 to 50 m water depths. The meiobenthic taxonomical richness at the Phyllophora field was higher and it was composed of 22 higher taxa (phylum, class, order) were found: Gromiida, Ciliophora, Foraminifera, Nematoda, Harpacticoida, Ostracoda, Acari, Kinorhyncha, Tardigrada, juvenile representatives of the macrobenthic fauna such as Polychaeta, Oligochaeta, Turbellaria, Nemertini, Bivalvia, Gastropoda, Cnidaria, Decapoda, Isopoda, Amphipoda, Cumacea, Ophiuridea, and

Chironomida. Nematoda had the highest share in abundance (42.3–93.9%). In the last 20 years, there has been a significant change in the composition of meiobenthos at the *Phyllophora* field and the aforementioned Foraminifera + Nematoda complex turned into a Nematoda + Harpacticoida complex [48].

During the researches a total of 22 higher taxa of meiobenthos in the Karkynitsky Bay and Phyllophora field of Zernov were presented. We have also recorded an increase in the average abundance of meiobenthos from 760 thous. ind.•m^{-2} in 1990s to 820 thous. ind.•m^{-2} in 2010–2011. Distribution of meiobenthos in the Phyllophora fields as in Karkinitsky Bay is determined by the following factors: the presence of oxygen deficiency, state of the *phyllophora* population, the composition and density of macrobenthos.

Nematoda dominated in meiobenthos of these two areas and their distribution was not homogenous. Total number of nematode species in the Karkynitsky Gulf and the *Phyllophora* Fields are 115 and 173 species, respectively. 8 species (*Paralinhomoeus deconincki, Metalinhomoeus zoterae, Leptolaimus* cf. *steineri*, *Sabatieria* sp A, *S. abyssalis*, *S. pulchra*, *Diplopeltus onustus* and *Terschellingia longicaudata*) with amphid abnormalities were found at seven stations. Among these species, individuals of females and juveniles with 4 and 3 amphids of species-specific form were isolated.

North-Eastern Part of the Black Sea Area Near the Kerch Strait

Part of the Black Sea near Kerch was studied in the depth range of 60–191 m. It is located at the north-eastern part of the Black Sea within the structures promising for oil and gas production. The bottom sediments are mostly represented by clayey and silty muds with a greater or lesser admixture of mollusc shells, ostracods and diatoms. Several stations were noteworthy by the smell of hydrogen sulfide. Previous studies conducted at this area (2005) have revealed low density and species richness of zoobenthos [49]. According to the authors, this could be due to the increase of methane, hydrogen sulfide and other gases in sediments. Further studies at the water area (2013) of this geodynamic activity zone resulted in a similar specificity of macrobenthic structure [50].

17 higher level of meiobenthic taxa were found at the first studied depth range of 60 - 249 m: Gromida, Ciliophora, Foraminifera, Cnidaria, Nematoda, Polychaeta, Turbellaria, Oligochaeta, Kinorhyncha, Nemertini, Bivalvia, Gastropoda, Harpacticoida, Ostracoda, Decapoda, Acari and Cumacea. Changes in the distribution of meiobenthic taxa and their population density

are determined by the increase in depth, as well as the properties of the habitat. Significant differences in the structure of meiobenthos were observed at the same depths in different parts of the investigated area. Nematodes were dominant. Species richness of nematodes was low and did not exceed 7–31 representatives at the stations [51].

Abnormal specimens of four nematode species (*S. abyssalis, S. pulchra, Sph. dispar, Sph. macrocirculus* and *Diplopeltoides* sp.) were prominent at seven stations. Abnormality was mainly in the non-standard number of amphids and we observed one or three and four elements instead of two (Figure 5). Additionally, one male specimen of *S. abyssalis* had teratic spiral elements. The share of abnormal individuals in the population is 5.0–100% in this area. Obviously, species composition was not rich and the presence of abnormalities was not diverse due to their habitat under actively geodynamic conditions. Anomalies were mainly in the non-standard number of amphids (Table 4).

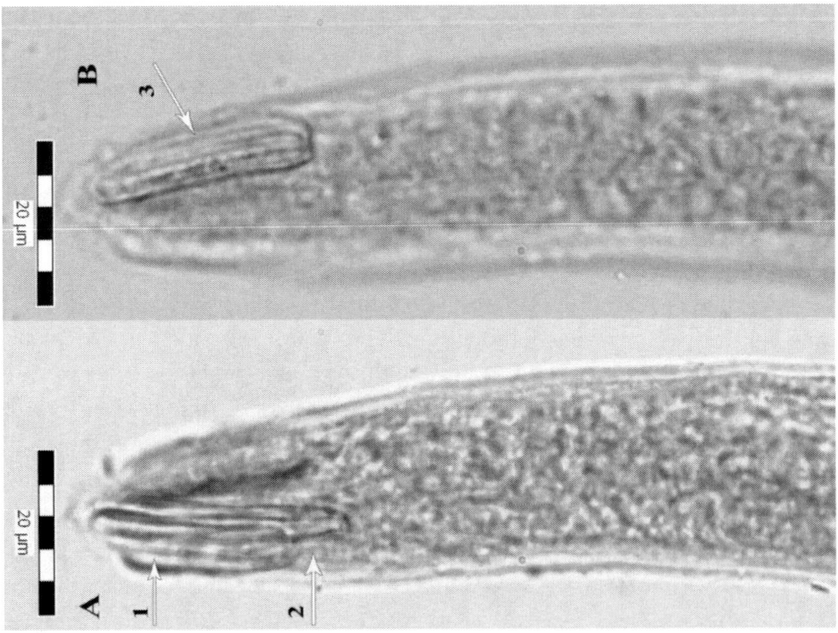

Figure 5. *Diplopeltoides* sp. with three elements of amphid in species-specific form found in area near the Kerch Strait (2013).

Table 4. Share of abnormal and normal individuals in the populations of nematode species found at the North-eastern part of the Black Sea area near Kerch Strait (2013)

Species	Number of amphids	Station/Depth, m													
		10/60		18/94		41/90		45/77		50/79		53/90		58/79	
		ind.	%	ind.	%	ind.	%	ind.	%	ind.	%	ind.	%	ind.	%
S. abyssalis	4 change 2*	- 1♂ 8	- 10.0 90.0	-	-	-	-	-	-	3♀ - 22	12.0 - 88.0	-	-	-	-
S. pulchra	4 2*	-	-	1♀ 11	8.3 91.7	1juv,1♀ 4	25.0 75.0	1♀ 8	11.1 88.9	-	-	-	-	-	-
Sph. dispar	4 1 2*	-	-	1♀ - 16	5.9 - 94.1	-	-	-	-	-	-	-	-	- 2juv 2	- 50.0 50.0
Sph. macrocirculus	1 2*	1♀ 0	100.0 0												
Diplopeltoides sp	3 2*	-	-	-	-	-	-	-	-	-	-	1♀ 0	100.0 -0	-	-

*Note: normal species-specific form and number of amphids.

The Bays of Sevastopol

The Bays of Sevastopol are located close to the south-western cape of the Crimean peninsula. The Sevastopolskaya Bay is one of the rare natural inland harbors on the coasts of Crimea. This bay has been under the effect of very heavy anthropogenic pollution for several decades, though it is an area that is vitally valuable for recreation and inhabitation. Sevastopol Bays included a number of natural inland harbors with inherent environment and contamination degree.

Sevastopolskaya Bay, Yuzhnaya and Balaklavskaya Bays are the most industrially loaded areas [52, 53 and 54]. In the sediments of these bays are significant concentrations of petroleum hydrocarbons and heavy metals. In addition, weak hydrodynamics determines the existence of the local sites with permanent hypoxia.

The area of Sevastopolskaya Bay is of an estuarial type with a length of approximately 7 km and its average width is 1 km. Average depth is 12 m, with a maximum of 22 m. This bay which is several hundred meters wide in its seaward part goes inland for about 8 km providing excellent conditions for ship docking, harboring and other maritime activities. During construction in 1976–1977, width of the protective breakwater was narrowed from 940 m to 550 m, which resulted in a reduction in the intensity of water exchange which was almost doubled.

The bay is actually a reservoir that receives industrial and domestic waste water and storm water from the catchment area. Increased concentrations of organic carbon in the bottom sediments result in anoxic/sulfidic conditions and support the flux of dissolved sulfides from sediments to the bottom layer of water. A consequence of this long-term industrial, military and municipal activity was the deterioration of marine environment of the bay and occurrence of expanded bottom areas that became contaminated with various sediment-associated pollutants including polycyclic aromatic hydrocarbons (PAHs), polychlorinated biphenyls (PCBs), pesticides, and metals. Apart from the fact that the Bay is heavily used by ships, the contribution to pollution of the Bay is made by domestic and urban run-off, which here is caused by intensive eutrophication, and hypertrophy stage of some of its regions (*Yuzhnaya* Bay) [55]. The status of the waters of the Bay is often accompanied by development of hypoxia, accumulation of organic carbon in bottom sediments and appearance of free hydrogen sulfide and other sulfur compounds recovered in the upper sediment layer and the bottom layer of water [56]. The discharge of industrial wastes into the Bay has led to the accumulation of various metals in sediments [57].

Extreme hypoxia is a regular feature of the inner part of the bay in summer. Moreover, sediments have become sulfidic and now serve as a source of hydrogen sulfide for the bottom layer of water.

The *outer part of the bay* is rather clean. This part represents the paleo riverbed of the same river that currently releases its waters in the head of Bay. Intensive alongshore currents effectively mix and transport pollutants off the coastal sources and the bay. Hypoxic events have never been reported for the waters of the outer part of the Sevastopol Bay. Still, reducing conditions have been recently reported for the bottom sediments from this site that accumulates particulate organic carbon. The outer bay can serve as a reference site with rather natural conditions in the water column and with potentially hypoxic conditions in the upper bottom sediments. Hypoxia and anoxia in the bottom layer of the Bay water in summertime is a result of a number of physical, biological, biogeochemical, and geochemical processes.

The response of meiobenthos to the anthropogenic impacts of different origin in the Sevastopol bays was studied during 2001–2013.

Sevastopolskaya Bay

Meiobenthos of Sevastopolskaya Bay was represented by 16 higher level taxa in different years: Gromiida, Ciliophora, Foraminifera, Nematoda, Kinorhyncha, Harpacticoida, Ostracoda, Acari, Tardigrada, juvenile Cnidaria, Polychaeta, Oligochaeta, Turbellaria, Nemertini, Bivalvia and Gastropoda. The dominant groups were Nematoda, Foraminifera (soft-shelled forms), Ciliophora and Harpacticoida. Minimum abundance was observed in the inner part of the bay. Meiobenthos included 7–12 groups of organisms at the studied stations [58, 59].

Species composition of nematodes was changing during the year. For example, the number of species varied from 7 to 39 at the same station under permanent hypoxia.

In this area, registration of abnormal individuals in the populations of several species (*S. pulchra, A. ponticus* and *A. setosus*) is no exception. Accordingly, Tables 5 and 5–1 show the contribution of abnormal individuals to the different populations of nematodes found at the Sevastopolskaya Bay in 2009 and 2013.

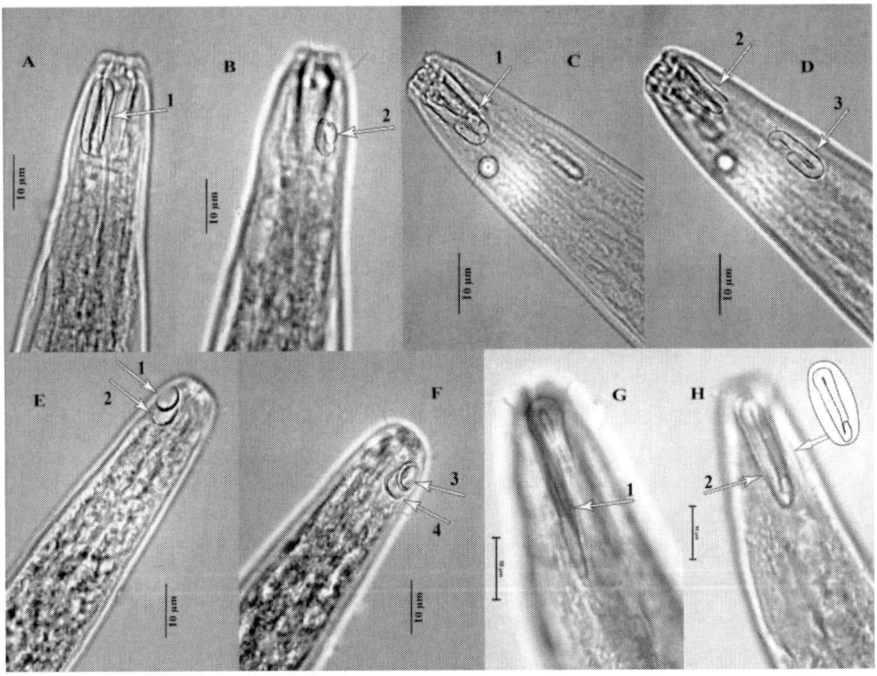

Figure 6. Varieties in abnormal development of amphids in some nematodes species found at the Sevastopolskaya Bay: A–D *Axonolaimus setosus:* A–B, amphids of one specimen with two elements of different forms: A – species-specific form, B – underdeveloped in oval form; C–D, one specimen with three different elements of amphid: one element normal and the two others are not in species-specific form and with abnormal topography on head; E–F, *Terschellingia longicaudata*: one specimen with doubling of the species-specific amphids. G–H, *Pseudolella* sp. a specimen with two elements of different forms: G, amphid of species-specific form, H, abnormal form.

Abnormal mature individuals were found in the populations of 11 species. Along with the standard number of elements, abnormal specimens had only one element of amphid or 3- 4 elements. In addition, disturbance in the structure of amphids, as far as until the loss of species-specific form was observed in some individuals of four species (Tables 5 and 5–1, Figure 6).

Apart from the fact that the Sevastopolskaya Bay is heavily used by ships, the contribution to the pollution of the Bay make domestic and urban run-off, which here is caused by intensive eutrophication, and some of its regions (bay Yuzhnaya) – hypertrophy stage [55]. This state of the waters of the Bay, often accompanied by development of hypoxia, the accumulation of organic carbon in bottom sediments of the Bay and the appearance of free hydrogen sulfide and other sulfur compounds recovered in the upper sediment layer and the bottom layer of water [56]. The discharge into the waters of the Bay of industrial wastes has led to the accumulation in sediments of various metals [57].

Extreme hypoxia is a regular feature of the Yuzhnaya bay in summertime. Moreover, sediments have become sulfidic and now serve as a source of hydrogen sulfide for the bottom layer of water.

Significant concentrations of petroleum hydrocarbons and heavy metals are found in the sediments of this bay. Additionally, weak hydrodynamics determines the existence of the local sites with permanent hypoxia. We studied seasonal changes in meiobenthic composition at the permanent hypoxic sites as a reaction to the varying concentrations of hydrogen sulfide in the environment between 2009 and 2010 in the framework of the 'Hypox' EC project as its [54]. A low species richness of nematodes and presence of abnormal specimens in three populations (*S. pulchra, P. quadristicha* and *M. kaurii*) were observed in different seasons. The spectrum of amphid abnormalities is repeatedly observed in other conditions considered above (Table 6).

The *outer part of the bay* is rather clean. This part represents the paleo riverbed of the same river that currently releases its waters to the head of the Bay. Intensive alongshore currents effectively mix and transport pollutants off the coastal sources and the bay. Hypoxic events have never been reported for the waters of the outer part of the Sevastopol Bay. Still, reducing conditions have been recently reported for the bottom sediments from this site that accumulates particulate organic carbon. The outer bay can serve as a reference site with rather natural conditions in the water column and with potentially hypoxic conditions in the upper bottom sediments. Hypoxia and anoxia in the bottom layer of the Bay water observed in summertime is a result of a number of physical, biological, biogeochemical, and geochemical processes.

As part of the same "Hypox" EC project, a parallel similar meiobenthos study was carried out at the control site (the Outer road of Sevastopol), which corresponds to a relatively clean environment with all its hydrochemical characteristics. So, any detection of the abnormalities was unexpected in the structure of nematode amphids at this area (Table 7). However, it is not surprising that these anomalies are found namely in those populations of species, mentioned above, for the other areas of the Black Sea.

Table 5. Share of abnormal and normal individuals in the populations of nematode species found at the Sevastopolskaya Bay (2009)

Species	Number of amphids	№ Station/Depth, m											
		1/3		11/10		13/11		15/17		18/9		19/14	
		ind.	%	ind.	%	ind.	%	ind.	%	ind.	%	ind.	%
A. setosus	4	-	-	-	-	-	-	1juv	3.3	-	-	1♀	1.2
	2 change							2♂1♀	10.0			1♀	1.2
	2*							26	86.7			76	97.6
Terschellingia cf. *longicaudata*	4	-	-	-	-	-	-	-	-	1♀	5.9	1juv	14.3
	2*									16	94.1	6	85.7
T. longicaudata	4	-	-	-	-	-	-	-	-	1juv1♂	100.0	-	-
	2*									0	0		
S. pulchra	4	1♀	6.7	1♀	11.2	2juv3♀	14.3	1juv	6.6	-	-	-	-
	2*	15	93.3	8	88.8	30	85.7	13	93.4				
Th. maeoticus	4	-	-	1♀	50.0	-	-	-	-	-	-	-	-
	2*			1	50.0								
Pseudolella sp.1	2 change	-	-	-	-	-	-	1♀	5.0	-	-	-	-
	2*							19	95.0				

*Note: normal species-specific form and number of amphids.

Table 5-1. Share of abnormal and normal individuals in the populations of nematode species found at the Sevastopolskaya Bay (May, 2013)

Species	Number of amphids	Station/Depth, m															
		3/8		8/14		9/10		12/14		14/14		16/16		19/14		30/17	
		ind.	%	ind.	%	ind.	%	ind.	%	ind.	%	ind.	%	ind.	%	ind.	%
S. pulchra	4	1♀	2.5	1♀	1.2	-	-	-	-	-	-	-	-	-	-	-	-
	2*	38	97.5	79	98.8												
A. ponticus	4	1♀	33.3	-	-	1♀	12.5	-	-	1♂	2.7	-	-	-	-	-	
	3	-	-			-	-			1♀	2.7			-	-		
	1change	-	-			-	-			-	-			1♂	1.6		-
	2*	2	66.7			7	87.5			34	94.6			60	98.4		
P. quadristicha	4	-	-	-	-	-	-	1♀	12.5	-	-	-	-	-	-	-	
	1change							-	-			1♀	14.2				-
	2*							7	87.5			6	85.8				
Sph. ostreae	4	-	-	-	-	-	-	-	-	-	-	-	-	-	-	-	
	1change											1♀	50.0				-
	2*											1	50.0				
Aponema sp.	3	-	-	-	-	-	-	-	-	1♂	50.0	-	-	-	-	-	-
	1									-	-					1♂	1.4
	2*									1	50.0					70	98.6
Molgolaimus sp.	1													1♀	5.5	-	-
	2*													17	94.5		

*Note: normal species-specific form and number of amphids.

Table 6. Share of abnormal and normal individuals in the populations of nematode species found at the Yuzhnaya bay in different seasons (2009–2010)

Species	Number of amphids	2009						2010			
		September		October		December		March		July	
		ind.	%	ind.	%	ind.	%	ind.	%	ind.	%
Sabatieria pulchra	4	1♀	6,6	1♀	1.6	-	-	1juv	33.4	-	-
	2*	14	93.4	62	98.4			2	66.6		
Parodontophora quadristicha	4		-	-	-	1♀	4.0	-	-	-	-
	2*	-				24	96.0			32	97.0
	2 change					-	-			1♀	3.0
Microlaimus kaurii	3	1juv	14.3	-	-	-	-	-	-	-	-
	2*	6	85.7								

*Note: normal species-specific form and number of amphids.

Table 7. Share of abnormal and normal individuals in the populations of nematode species found at the site of the Outer road of Sevastopol (2009–2010)

Species	Number of amphids	2009								2010		2013	
		July		September		October		December		March		May	
		ind.	%	ind.	%	ind.	%	ind.	%	ind.	%	ind.	%
Sabatieria pulchra	4	1juv,1♀	8.7	-	-	5♀	4.7	-	-	8♀	15.6	-	-
	2*	23	91.3			102	95.3			43	84.4		
Spirinia parasitifera	4	1♀	3.0	1♀	5.3	-	-			-	-	-	-
	3	-	-	-	-	1♀	6.7	-	-			1♀	1.1
	2*	33	97.0	19	94.7	15	93.3					93	98.9
Paracomesoma dubium	4	-	-	-	-	-	-	1♀	2.7	-	-	-	-
	2*							37	97.3				
Microlaimus kaurii	3	-	-	-	-	-	-	1♀	5.6	-	-	-	-
	2*							18	94.4				

*Note: normal species-specific form and number of amphids.

It should be noted that, the most significant thing in the populations of three species registered at this site, in our view, is a disturbance, namely the change in species-specific forms of amphids (Figure 7). 3 anomalous species was found in the Outer road of Sevastopol, among these representatives previously such disturbances no observed.

It can be assumed that the polluted bottom sediments and the inhabiting meiofauna as a result of storms and active traffic shipping arrive into adjacent areas from the Sevastopolskaya Bay.

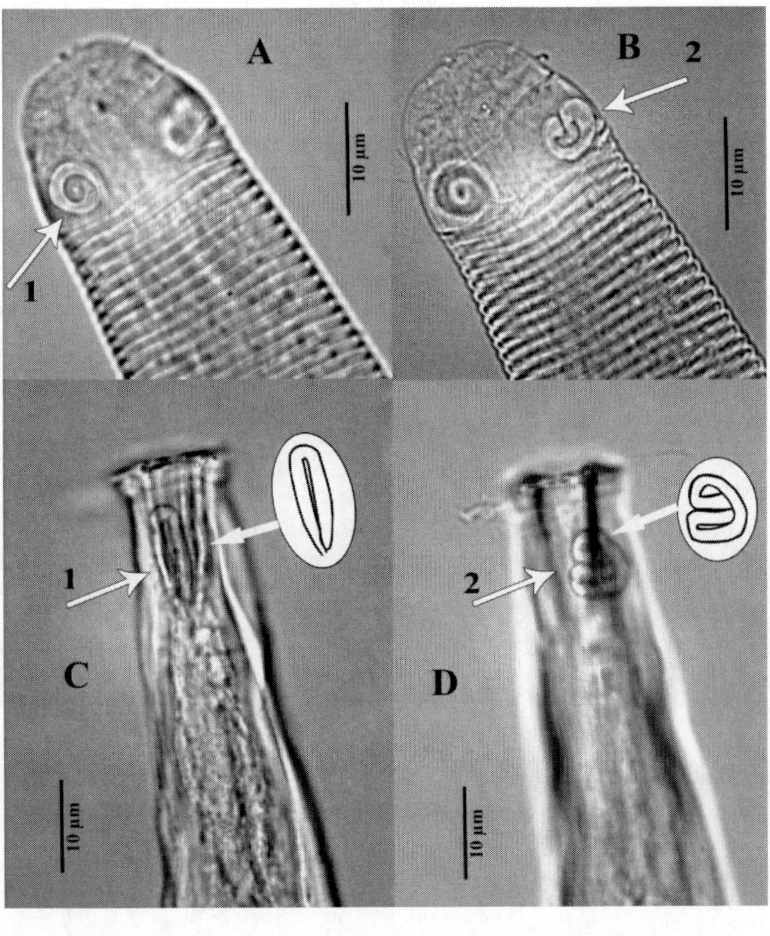

Figure 7. Abnormal developments of amphids in two nematode species from the Outer road of Sevastopol. **A-B** *Desmodora conica*, **C-D** *Axonolaimus setosus*. Specimens with two different elements of amphid: **A, C** – species-specific form, **B, D** – disturbance in the species-specific form.

The Kruglaya (Omega) Bay

The Kruglaya (Omega) Bay is a semi-enclosed basin situated in the south-western part of the Crimean Peninsula. It is part of a larger complex of bays that are located within the Sevastopol region, at the coastal line between Cape Chersones and Sevastopol Bay. The Kruglaya Bay covers an area of 0.66 km^2 (1.4 km long and 0.9 km wide). Water depths decrease gradually from 17 m at the entrance to 2 m in the most distant part of the bay.

Despite the fact that the Kruglaya Bay is located inside the area of Sevastopol city, no industrial objects are located at its coast. Instead, there are resorts, parks and apartment buildings found along its coast. Therefore conditions in Kruglaya Bay could be expected to represent pristine conditions. However, this is not the case instead, small water depths, a rather weak ventilation, and discharge of municipal waste waters result in local temporal hypoxic conditions at the bottom.

Local black fields and spots were noted at the bottom of some parts of the bay and they were attributed to the permanent hypoxic zones as a result of chemical analyses. Spot-like sites with dead macrophyte *Zostera marina* – zones and periodic sulfide contaminations were observed at bottom sediments

Kruglaya (Omega) and Kazachya Bays also belong to the most environmentally prosperous bays. In the framework of 'Hypox' and 'Perseus' EC projects, study of seasonal changes in meiobenthos was conducted at a local site in Kruglaya Bay where bottom sediments are characterized by permanently reduced conditions. The main reason for the periodic near-bottom hypoxia is the intensive use of the bay as a recreational area and the discharge of waste water. During the research of meiobenthos, changes of hypoxic conditions and hydrogen sulfide concentration were observed in sediments [54].

Meiobenthos under hypoxic conditions in Kruglaya Bay was represented by 17 taxa: Gromiida, Ciliophora, Foraminifera, Nematoda, Kinorhyncha, Harpacticoida, Ostracoda, Acari, Tardigrada, Gastrotricha and juvenile Cnidaria, Polychaeta, Oligochaeta, Turbellaria, Nemertini, Bivalvia, Gastropoda.

In Kruglaya Bay, the populations of four species of nematodes (*S. pulchra, Sp. parasitifera, P. dubium and M. kaurii*) showed developmental abnormalities in their amphids. Abnormality was reflected in the number of amphid elements which were out of the standard (3 and 4) (Table 8, Figure 8). Registeration of abnormal specimens was almost regular for the two species *S. pulchra* and *Sp. parasitifera*.

Table 8. Share of abnormal and normal individuals in the populations of nematode species found at the Kruglaya Bay during different seasons (St. 5, 2009-2013)

Species	Number of amphids	2009								2010		2013	
		July		September		October		December		March		May	
		ind.	%	ind.	%	ind.	%	ind.	%	ind.	%	ind.	%
S. pulchra	4	1juv1♀	8.7	-	-	5♀	4.7	-	-	8♀	15.6	-	-
	2*	23	91.3			102	95.3			43	84.4		
Spirinia parasitifera	4	1♀	3.0	1♀	5.3	-	-			-	-	-	-
	3	-	-	-	-	1♀	6.7	-	-			1♀	1.1
	2*	33	97.0	19	94.7	15	93.3					93	98.9
Paracomesoma dubium	4	-	-	-	-	-	-	1♀	2.7	-	-	-	-
	2*							37	97.3				
M. kaurii	3	-	-	-	-	-	-	1♀	5.6	-	-	-	-
	2*							18	94.4				

*Note: normal species-specific form and number of amphids.

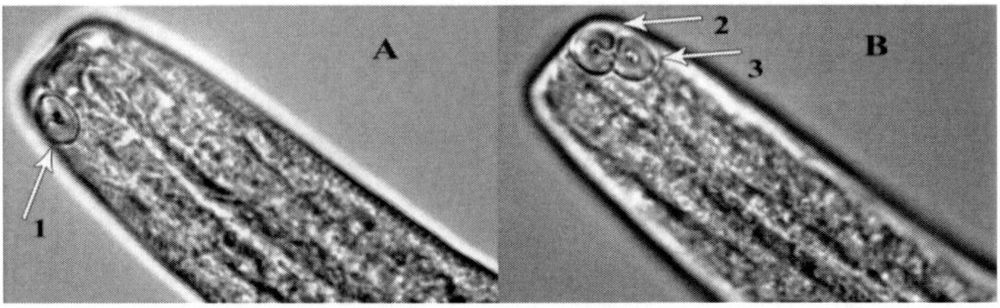

Figure 8. Abnormal development of amphids in one specimen of *Spirinia parasitifera* found at the Kruglaya Bay, **A** – one element of amphid species-specific form, **B** – double amphids consisting two elements, an abnormal and a species-specific form.

In studied area of the Kazachya Bay, the nematode fauna consisted of 54 species. In this Bay, abnormal juvenile and mature individuals were marked in the populations of six nematode species (Table 9). Two representatives (*Sphaerocephalum crassicauda* and *Monhystera* sp.) were also found along with the above-mentioned species (*S. pulchra, Sp. parasitifera, A. setosus, Th. sabulicola*).

Kruglaya and Kazachya Bays, and the waters of the Outer road of Sevastopol as well are the most favorables in terms of environmental conditions in Sevastopol region [60]. However, several nematode species with similar amphid abnormalities were also found in these areas.

Table 9. Share of abnormal and normal individuals in the populations of nematode species found at the Kazachya Bay (2003)

Species	Number of amphids	№ Station					
		1		2		3	
		ind.	%	ind.	%	ind.	%
Sabatieria pulchra	4 2*	-	-	-	-	1♀ 55	1.8 98.2
Spirinia parasitifera	4 2*	-	-	1juv 3	25.0 75.0	-	-
Axonolaimus setosus	4 2*	-	-	-	-	1♀ 19	5.0 95.0
Monhystera sp.	4 2*	-	-	1♀ 2	33.4 66.6	-	-
Theristus sabulicola	4 2*	1♀ 2	33.4 66.6	-	-	-	-
Sphaerolaimus crassicaudata	3 2*	1♂ 3	25.0 75.0	-	-	-	-

*Note: normal species-specific form and number of amphids.

Abnormalities of nematodes in the Kruglaya and Kazachya Bays were also limited to non-standard number (3–4) of amphid elements, whereas in the area of the Outer road of Sevastopol, changes in the forms of amphid elements were shown for all the abnormal species (see Table 7).

Balaklavskaya Bay

Bottom sediments of Balaklavskaya Bay, as well as Yuznaya bay, are characterized by a high level of technogenic pollution [52, 53]. One of the main sources of pollution (oil products, heavy metals, etc.) based in this area was the Navy for many decades. The content of petroleum hydrocarbons in the sediments of the bay is much higher in the background, and the maximum values exceed the acceptable limits by dozens of times. Heavy metal (Cd, Pb and Cu) concentrations are high locally. Bottom sediments in the inner part of the bay (stations 1–6) are represented by black silts with a noticeable smell of hydrogen sulfide. In recent years, full or partial demilitarization of this area have implied improvement of the quality of marine environment in the future.

Despite the high level of multicomponent anthropogenic pollution, the bottom sediments were inhabited by representatives of 7–14 taxa, and the meiobenthos density was in the range of 200 thous.–1.7 million ind./m^2. It is known as a rule that, pollution suppresses the development of benthic fauna and leads to destruction of the all the communities. In Balaklavskaya Bay, as in the Yuznaya bay, higher structural characteristics of meiobenthos were similar to those of more affluent areas or even exceed them. Obviously, this is due the favorable trophic conditions for bottom fauna on t one hand, and wide adaptive mechanisms of organisms on the other hand. Nematodes were found to be dominant in meiobenthos composition.

In Balaklavskaya Bay, the anomalies in the amphid development of four species (*S. pulchra, S. abyssalis, A. setosus* and *P. quadristicha*) were recorded at nine of the seventeen stations, which covered the inner part of the bay towards the exit of the sea in 2005 (Table 10). It should be noted that, abnormalities were found at station 9, where a minimum density of meiobenthos (94 thous.ind.•m^{-2}) was observed and, as well as at stations 2, 6 and 7, where a maximum density of meiobenthos (1.0 - 1.7 million ind.•m^{-2}) was recorded. It should also be considered that, an earlier (2001) record was noted for abnormal specimens belonging to 6 species including *S. pulchra* and *A. setosus* [32].

Table 10. Share of abnormal and normal individuals in the populations of nematode species found at the Balaklavskaya Bay (2005)

Species	Number of amphids	№ Station/Depth, m																	
		1/6		2/7		5/9		6/5		7/9.5		9/12		10/11		13/22		16/19	
		ind	%	ind	%	ind	%	ind	%	ind	%	ind	%	ind	%	ind	%	ind	%
S. pulchra	4	1♀	1.7	2♀	1.3	1♀	1.0	-	-	-	-	-	-	1♀	5.8	-	-	-	-
	2*	58	98.3	59.0	98.7	99.0	99.0							16	94.2				
S. abyssalis	4	-	-	-	-	-	-	-	-	-	-	-	-	-	-	1♀	5.2	-	-
	2*															18	94.8		
A. setosus	4	-	-	-	-	1♀	4.4	1♀	3.4	2♀	4.6	-	-	1♀	25.0	2♀3 juv	14.7	2juv	2.9
	2 change					-	-	-	-	1♀	2.3			-	-	-	-	-	-
	2*					22	95.6	2	66.6	40	93.1			3	75.0	29	85.3	65	97.1
P. quadristicha	4	-	-	-	-	-	-	-	-	1♀	9.1	-	-	-	-	-	-	-	-
	2 change									-	-	1♂	11.2						
	2*									10	90.9	8	88.8						

*Note: normal species-specific form and number of amphids.

Sivash Lake

Sivash is a gulf located at the western part of the Sea of Azov and separates the Crimean peninsula from the mainland. Although Sivash is a fairly large water basin area of 2453 km^2, its depth range is small (0.5–1.5 m) and the maximum depth is only 3.2 m. In summer, water gets strongly warmer and emits a putrid smell, and that's why Sivash is called the Dead or Rotten Sea. Compounds in brine Sivash Lake include chlorine, sodium, potassium and magnesium, magnesium bromide, magnesium sulfate and other salts.

In April 2014, Ukraine stopped the supply of Dnieper river water in the North Crimean Canal, resulting in a salinity increase in the Gulf of Sivash (the Sea of Azov). During our surveys in 2013, salinity was found to be 10–40 g/L [61]. In 2015, the concentration of salinity was two-three times higher than in 2013.

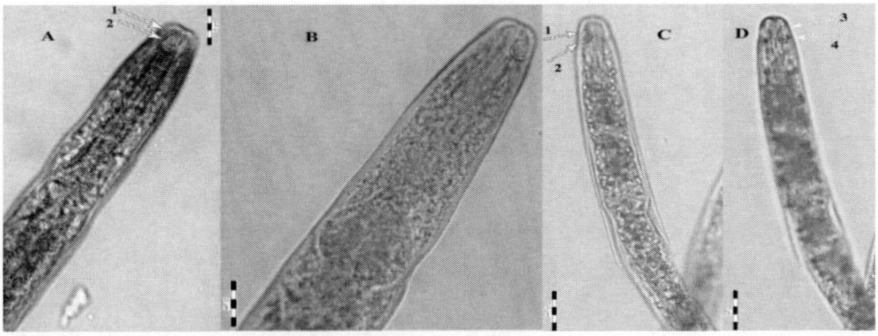

Figure 9. *Terschellingia longicaudata* with double amphids composed of elements in a species-specific form found in the Sivash Lake; A- one specimen with doubling of the species-specific amphids, B- one specimen with the normal amphid, C-D one specimen with doubling of the species-specific amphid.

The taxonomic structure of meiobenthos showed significant changes until 2015. Species composition of free-living nematodes was not diverse in both study periods and the distribution of density was uneven. Since these observations were unfortunately limited to the coastal area at 4–6 sites, we have not still completed the information about the richness of fauna in this lake. However, the study revealed similar abnormalities in five nematode populations (*S. pulchra, Monhystera* sp., *T. longicaudata, T. cf. longicaudata…* and *Sph. gracilis*) (Table 11, Figure 9). Abnormality observed in all specimens was the presence of four elements of amphids. One specimen of *Sph. gracilis* had four amphid elements with a species-specific form and size, but one pair was located at the standard (lateral) part of the body, and the other pair lwas located on the dorso-ventral. Contribution of abnormal individuals to the populations ranged from 0.5% to 90.0%.

Table 11. Share of abnormal and normal individuals in the populations of nematode species from the Sivash Lake

Species	Number of amphids	2013		2015					
		St. 1-2		St. 1		St. 3		St. 5	
		ind.	%	ind.	%	ind.	%	ind.	%
S. pulchra	4 2*	1♀ 9	10.0 90.0	-	-	-	-	-	-
Monhystera sp. A	4 2*	1♀ 6	14,3 85.7	-	-	-	-	-	-
Terschellingia cf. longicaudata	4 2*	-	-	1♀ 182	0.5 99.5	-	-	2juv 150	1.3 98.7
T. longicaudata	4 2*	-	-	-	-	5♀ 97	5.0 95.0	-	-
Sph. gracilis	4 2*	-	-	-	-	-	-	1juv 9	10.0 90.0

*Note: normal species-specific form and number of amphids.

The Shelf and Upper Slope of the Turkish Areas

The shelf and slope area of the Black Sea off the İstanbul Strait is characterized by the inflow of warm, saline Mediterranean Water and the rim current. These waters are the carrier of oxygen to the deep basin. It flows over 60 m sill in the Bosporus, follows a submarine channel towards NE over the inner shelf, and reaches the mid-shelf region where it spreads over the seafloor to form a 2–3 m thick sheet [62]. It mixes with cold intermediate water between depths of 50 and 75 m, cools to 8°C at the shelf break, becomes denser and cascades down the slope to form intermediate depth intrusions. Maximum depth of penetration of the MW is believed to be – 500 m [63].

Water exchange occurs in the Black Sea through the Bosporus, with Black Sea water flowing as an upper current through the Bosporus, while a lower current transports saline water from the Sea of Marmara to the Black Sea. This dense, saline water sinks at the continental slope and drives convection.

In the Istanbul Strait's (Bosporus) outlet area of the Black Sea, the less saline surface waters of the Black Sea interact with the saline Mediterranean waters, creating a special ecological system and rapid transitions from oxic, hypoxic and anoxic water conditions [62, 64].

The meiobenthos of the depths between 75 and 300 m included 21 higher taxa. At present, our study of the species composition of nematodes from the Bosporus area is not finished. However, according to the preliminary data based on two studies (2009-2010), the nematode fauna includes at least 200 species in this area. In 2009, abnormal specimens of three populations were found in the community of free-living nematodes at three stations of among nine (cruise RV 'Arar') at the Istanbul Strait's (Bosporus) outlet of the Black Sea along the oxic/anoxic interface (Table 12, Figure 10). Distinct morphological abnormalities in amphids of the specimens belonging to three species populations (*Sabatieria* sp. A, *S. praebosporica* and *Terschellingia* cf. *longicaudata*) were detected at three stations in the depth range of 75– 174 m. Species with deviations in amphids were distributed at depths ranging from 75–250 m from the surface to a depth of 4 cm in the bottom sediment column. In *Terschellingia* cf. *longicaudata,* one specimen was found to have three elements instead of two elements of amphids; one specimen of *S. praebosporica* had four elements instead of two elements of amphid. Figure 5 demonstrates variations of disturbances not only in the number of amphids (3 and 4) which are in a different structure from the normal, but also in the location of the amphids relative to the axis of body symmetry.

In this area, specimens with distinct amphid abnormalities were recorded in other three species populations (*Paracanthonchus caecus, Sabatieria abyssalis and S. pulchra*) obtained at six stations in the depth range of 81– 174 m in 2010 (cruise RV 'Maria S. Merian'). These collections include specimens of nematodes with abnormal chemoreceptor organs in the form of double species-specific elements which are arranged one above the other (Table 12–1). However, only one of the two registered individuals of *P. caecus* and one among the 19 specimens of *S. pulchra* had four-amphids recorded at a depth of 179 m. In the populations of *S. abyssalis*, the share of individuals with similar abnormal amphid development was 7.1–25% at a depth of 81–250 m.

In conclusion, the amphid abnormalities observed in the Bosporus Strait were analogical to those found in environments with high technogenic impact and in the NW areas with methane gas seeps.

At present, almost any water body is subject to human impact. The main sources of pollution of coastal waters and sediments of urbanized areas are domestic waste water and various technological components. The negative effect is revealed in extinction of species, community perturbations, serious physiological changes arising in organisms, resulting in different pathological phenomena. We can assume that climate change also plays a role in these processes.

Thousands of tons of chemicals enter the Black Sea and interact with each other, transforming into other compounds, often with more toxic properties under the influence of physical, biological, and photochemical oxidation processes. However, the known chemical methods do not allow analysis of the whole complex of factors, which the impact is experiencing a particular body of water, taking into account the effects of each and collectively, on the individual links of the ecosystem. Existing methods for the quantitative determination of chemical compounds in natural waters are used only for a limited list, which creates considerable difficulties in the environmental assessment of the aquatic environment.

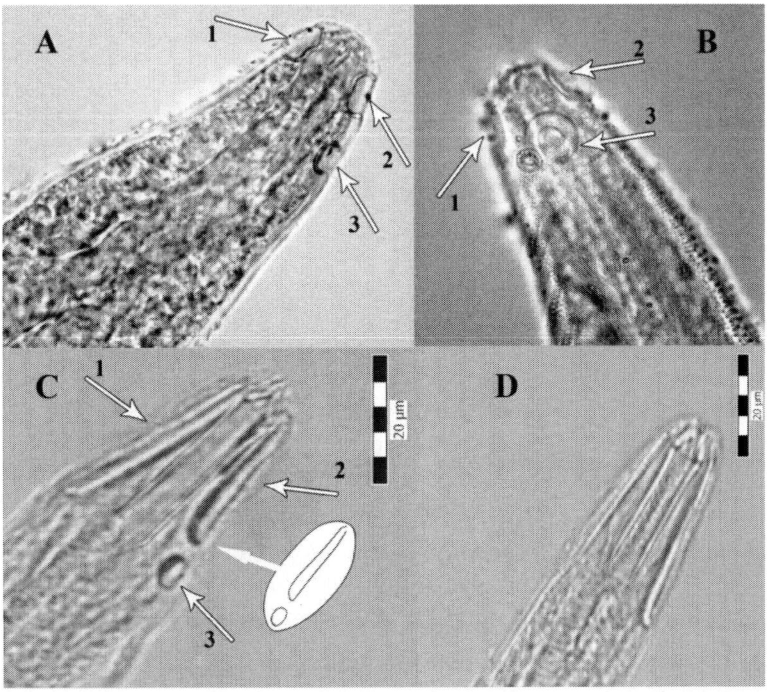

Figure 10. Variety in abnormal development of amphids in some species of nematodes found along the Oxic/Anoxic interface at the Bosporus area of the Black Sea: A - *Terschellingia* cf. *longicaudata*, individual has three elements instead of two elements of amphid; B - *Sabatieria praebosporica*, individual has three elements of abnormal topography instead of two elements of amphid; C-D – *Parodontophora* cf. *quadristicha:* C – abnormal individual, D- normal individual.

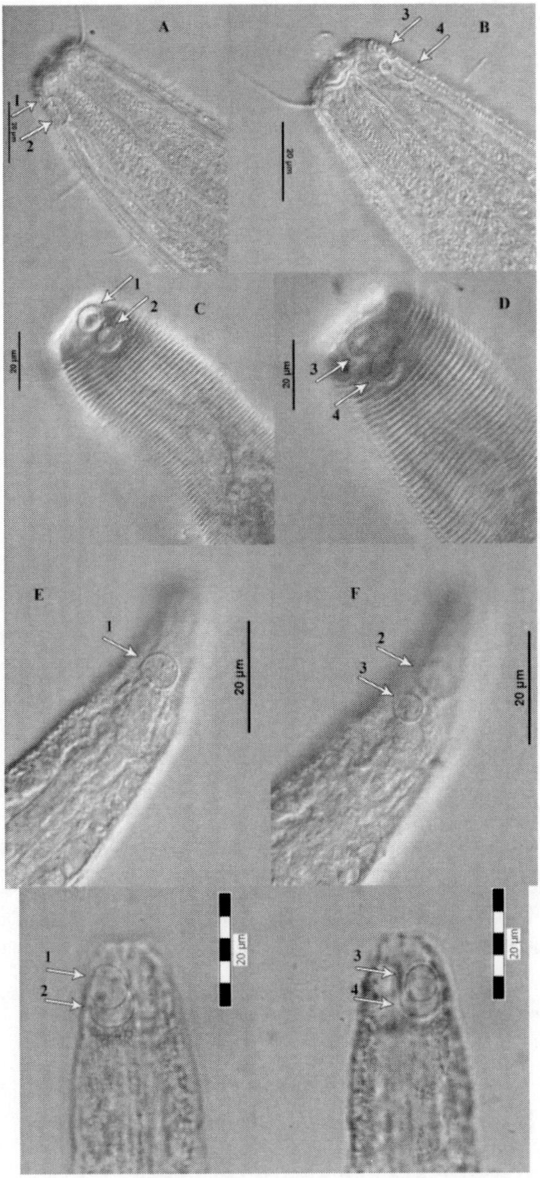

Figure 11. Variety in abnormal amphids of some nematode species found in Sinop Bay (A-F) and the Rize (lower row) coastal area (Southern part of the Black Sea, Turkey). A-B one specimen *Sabatieria* sp. with four elements of amphid; C-D one specimen *Desmodora pontica* with four elements of amphid; E-F one specimen *Microlaimus* sp with three elements of amphid; lower row -*Terschellingia longicaudata* with doubling of the species-specific amphid.

Table 12. Share of abnormal and normal individuals in the populations ofnematode species in the column of bottom sediments (cm) along Oxic/ Anoxic interface at the Istanbul Strait's (Bosporus) outlet area of the Black Sea (RV '*Arar*', 2009)

Species	Number of amphids	№ Station/Depth, m							
		1/75		4/80				9/250	
		Horizon, cm							
		0-1		1-2		2-3		3-4	
		ind.	%	ind.	%	ind.	%	ind.	%
Sabatieria sp. A	4	-	-	-	-	-	-	1♂2♀	5.1
	2*							56	94.9
S. praebosporica	4	1juv	1.0	-	-	-	-	-	-
	2*	99	99.0						
T. longicaudata	3	-	-	1♀	4.0	1♂	12.5	-	-
	2*			24	96.0	7	87.5		

*Note: normal species-specific form and number of amphids.

Long-term investigations of chromosome mutagenesis in dominant aquatic species of the Black Sea [65] have shown an increasing number of chromosome mutations in organisms coming from regions of increased mutagenic concentrations such as xenobiotics, mercury and PCB.

In 1978, anomalous specimens of *Nereis succinea* (Polychaeta) lacking the V group of paragnaths were found for the first time near the mouth of the Danube. In 1982–1983, such specimens were recorded to comprise 10 to 100% of the total population in a particular area [66].

Table 12-1. Share of abnormal and normal individuals in the populations of nematode species along Oxic/Anoxic interface at the Istanbul Strait's (Bosporus) outlet area of the Black Sea (RV *'Maria S. Merian'*, 2010)

Species	Number of amphids	Station/Depth, m									
		228/174		235/159		243/153		283/117		298/81	
		ind.	%	ind.	%	ind.	%	ind.	%	ind.	%
S. pulchra	4	1juv	5.2	-	-	-	-	2juv	33.3	-	-
	2*	18	94.8					4	66.7		
S. abyssalis	4	1♀	100.0	2juv	5.2	1juv	14.3	-	-	1♀	7.1
	2*	0	0	36	94.8	6	85.7			13	92.9
Sph. gracilis	4	-	-	-	-	1juv1♀	22.2	-	-	-	-
	2*					7	77.8				
P. caecus	4	1♀	50.0								
	2*	1	50.0								

*Note: normal species-specific form and number of amphids.

Benthic organisms, including the free-living nematodes, living in non-uniform, relatively constant conditions are continuous monitors for the quality of the environment. They summarize the effects of complicated chemicals and pollutants accumulated in bottom sediments, and in a certain way they react to changes in living conditions. Since the free-living nematodes are inhabitants of sediment, they are constantly exposed to the compounds in their environment for their entire life cycle. Dissolved forms of pollutants in the pore water are absorbed directly by the body of nematodes and substances in the form of suspended particles and colloids enter the body in the course of grazing detritus, waste products of other organisms released are through their digestive tract, where a partial detoxification occurs, followed by deposition of the remaining material in the body of the individual.

Rightly pointed out by several authors [67–69], it is not enough to observe sediment pollution only by chemical methods. Methods of monitoring the state of the environment on the basis of the results of chemical analysis of certain chemicals that are not related to the observation of a biological effect are not indisputable. Chemical analyzes can provide information about the defined pollution concentration, but the toxicity of chemical compounds depends not only on their concentration, but primarily on their chemical form and on individual susceptibility of benthic animals. Contaminated sediments contain toxic contaminants, many of which have a lethal effect on the body in a short time (heavy metals, chlorophenol) or long-term damaging effects, including reproduction disorders (DOM, dioxins) and cause mutations of benthic organisms. Genetic analysis of various adaptations aquatic taxa from bacteria to fish [70] indicates the increased resistance of some organisms to contamination with heavy metals (Zn, Cd, Cu, Co, Cr and Hg). However, the reasons for this phenomenon are not always obvious.

The results of previous studies [71] have shown that the eurybiont nematode species *Metoncholaimus demani* and *Sabatieria pulchra* are resistant to contamination by chloroform soluble bitumen in the Black Sea. These species are found to be resistant to oil contents up to 5g/100g of sediment in the environment. Two other species, *T. longicaudata* and *Th. sabulicola* are also adapted to such life conditions, however to a lesser extent. It is obvious that the same type of violation of vital organs in the same nematode species is a result of chromosomal mutagenesis. Unfortunately, it is not known that substances cause these mutations. In present we can only talk about the impact on nematode multicomponent natural and antropogenic factors and characteristic reaction of these animals on negative change their habitat. Many widespread species of nematodes at the Black Sea had similar anomalies.

Described developmental disorders in amphids of the nematodes wich were found near the Bulgarian continental shelf do not affect the form of amphids, whereas the nematodes from the Donuzlav Lake, Sevastopol Bays and NW part of the Black Sea shown dramatic changes in species-specific forms of these chemoreceptor organs.

With the new data, the number of nematode species, which have been subjected to deviations increased to 28 valid species and 21 morphotypes in the Black Sea (see Table 1). Abnormal specimens are very often found in the of populations of *Terschellingia longicaudata* (0.6–100%), *Axonolaimus setosus* (2.4–100%), *Parodontophora quadristicha* (2.9–100%), *Sabatieria pulchra* (1.6–33.4%), *S. abyssalis* (6.7–16.7%), and *Spirinia parasitifera* (5.0–25.0%).

We presented here only some of the available data regarding morphological abnormalities of amphids. Abnormal morphology of these organs can undoubtedly make changes in their function, which mainly is the function of chemoreception according to many authors. On the other hand, perhaps the disfunction of organs determines the variability of their structure and their number, as well as their localization in the body.

These results, based on previous and new data, have shown that a spectrum of anomalies in the development of nematode amphids is not a random phenomenon in the Black Sea. Most often they are found in areas with high anthropogenic load and natural extreme environments. It can be assumed that a constant or intermittent exposure to extreme environmental conditions may contribute to the development of various adaptation mechanisms in nematodes.

Depending on the thickness of sediment layer in which they live, nematodes use the available life space and resources at maximum. They colonize even the lower layers of bottom sediments despite the depletion in dissolved oxygen. Throughout their life cycle, they constantly feel the impact of each pollutant, and their compounds formed in the environment. High resistance of free-living nematodes to extreme conditions (heavy metals and petroleum products) has been shown by several researchers [31, 71–77]. It is known that some species of nematodes are able to withstand high concentrations of hydrogen sulfide in the ground due to the adaptation mechanisms of these species to toxic substances [78–80].

Based on these results, it can be assumed that toxic habitats may cause mutagenic effect in the formation of chemoreceptor organs and physiological correction on certain stages of ontogenesis. Abnormalities in nematodes can perhaps be hereditary, because they are found in females, males and juveniles of the same species. Prolonged exposure to heavy metals may lead to physiological acclimatization in animals or to changes in their behavior, determining natural selection of individuals with a high resistance, leading to genetic adaptation, which will be inherited [81].

Post-embryonic development is accompanied by periodic molting of free-living nematodes, i.e., juvenile individuals in the external environment passes through 4 stages of molting. The larvae of stages I and II are those known to develop trophic-sensory and caudal parts of the body [82, 83]. Perhaps the appearance of anomalies in the development of amphids is formed between the stages I and II when 4 fovea are recorded [84].

Thus, the spectrum of anomalies in the nematode species of the Black Sea are as follows: 1) the number of amphids may be 1–5 instead of 2 – normal; 2) The individual may have two amphids, but in different shapes; 3) one of the two amphids in an individual can be underdeveloped; 4) one of the amphids can be dissected into parts; 5) Location of the amphids with respect to the front edge and the axis of symmetry may be violated.

Based on the results, we can assume that the presence of local zones of extreme conditions in the environment led to a violation of the process of formation of lateral organs in ontogenesis. If the intensity of exposure to the environment caused by such phenomena is sufficiently large and long-term, the process of natural selection in genotypic composition of the population may continue in the way of polymerization or oligomerization.

Variability in chemoreceptor organs is apparently due to the adaptive capacity of the individuals in the population. These deviations may indicate plasticity and possibly cause appearance of modifications in species. However, further special studies are needed to determine if any evolutionary prospects are present in these structural abnormalities in the chemoreceptor organs of nematodes.

CONCLUSION

In conclusion, it should be noted in the present work that, we don't aim to explain this phenomenon from the standpoint of modern science and its role in the evolution of free-living nematodes and biological systems. Having accumulated a database of long-term data about the manifestation of the abnormal biological phenomena across the Black Sea, we aim to make it accessible for scientists from different fields of knowledge in biology and to draw attention for further study describing an extraordinary aspect of the phenotypic, and possibly genetic variability of free-living nematodes - the most numerous and important component in marine benthic ecosystems.

Given the complexity of the problem and the lack of any information about this amazing phenomenon in other basins, we encourage to join efforts of scientists from different countries for the knowledge and undestanding of biological effects, for example, marine nematodes, which stimulate the mechanisms of adaptation of biosystems to environmental conditions and its significance in progressive evolution. We consider the Black Sea as a unique natural laboratory for the development of original approaches in solving these problems.

The anthropogenic impact on the Black Sea ecosystem is steadily increasing. The negative effect is revealed in the extinction of species, community perturbations, serious physiological changes arising in organisms, resulting in different pathological phenomena.

We should underline that similarity and duplication of disorders in the number of amphid elements in the same species was found not only in the regions under anthropogenic (technogenic) impacts. We found the analogical phenomena in natural extreme conditions (methane seeps, permanent hydrogen sulfide conditions, hypersalinity etc.) and in areas with a relatively normal ecology (see Figures 2, 4, 6, 9 and 11).

The genetic apparatus is not only a tool of transmission and characteristics from generation to generation, but also a kind of regulator of metabolic processes throughout the life of the organism [86]. Pollution, in particular the different chemical compounds that fall into the environment, may be the factors of mutational pressure on the environment. Determination of mutagens requires expensive and time-consuming experiments. Moreover, assessment of genetic activity requires not only the study of the individual elements and compounds, but also their innumerable combinations. Despite the widespread development worldwide for the research to identify environmental mutagens, the information about the genetic effects of chemical compounds on marine animals is a small percentage.

We should underline that similarity and duplication of disorders in the number of amphid elements in the same species was found not only in the regions under anthropogenic (technogenic) impacts. The same phenomena were found to be marked in areas with natural extreme conditions (methane seeps, permanent hydrogen sulfide conditions).

The genetic apparatus is not only a tool of transmission and characteristics from generation to generation, but also a kind of regulator of metabolic processes throughout the life of the organism [85]. Pollution, in particular the different chemical compounds that fall into the environment, may be a factor of mutational pressure in the environment. Determination of mutagens requires expensive and time-consuming experiments. Despite the widespread development worldwide research to identify environmental mutagens, to date, the genetic effects of a small percentage of chemical compounds have been studied. Moreover, assessment of genetic activity requires not only the individual compounds, but also their innumerable combinations.

Repeatability of the described abnormalities in the Black Sea confirms the indisputable authenticity of our long-term observations and gives us reason to summarize the previous and new data to generalize the findings. Traditionally, three main levels of replications are recognized; exact, partial and conceptual. The first level, exact (sometimes also known as 'direct' replication provides the highest fidelity [86, 87].

Thus, we should recognize that we are witnessing the 'direct' replication of the phenomenon, in nature, but not in experiment. The consequence of this may be the appearance of new nematode species with acquired described morphological features.

We can assume that nematodes with amphid abnormalities are more widespread in the Black Sea. But, due to the fact that many of the nematodes do not have such distinct amphids as the species mentioned above, they are not registered.

Having considered the change in body of nematodes, and correlating observed results with the theory of evolutionary process of animals A.N. Severtsov's [88], we conclude that we have every reason presumably talk about aromorphoses of free-living nematodes species or idioadaptations in different genera. Obviously, these two directions of adaptive evolution in nematodes, suggest the formation of new species at this level of the organization. In particular, it is probable to make morpho-physiological differentiation of abnormal individuals at the level of the next generations, and the selection of individual branches with specific organization of amphids associated with the original species and genera.

Most researchers [89–91] distinguish two stages of evolution: microevolution (the emergence of new species) and macroevolution (the subsequent emergence of larger taxa up to the type).

Microevolution is a process in populations that and with the formation of new species. Obviously, we witness the nascent process of microevolution of nematode fauna in the Black Sea.

Currently, similar abnormalities have a wide distribution in the Black Sea for 28 species and 21 morphotypes of nematodes. It is obvious that the same type of abnormality of vital organs is a result of chromosomal mutagenesis. It is not known which substances caused these mutations. We can only talk about the permanent impact of multicomponent natural factors on nematodes and anthropogenically chemical contaminations in habitats and about characteristic reactions of these organisms. Described developmental disorders in amphids of nematodes which were found near the Bulgarian continental shelf do not affect the form of the amphids, whereas specimens with clear changes of species-specific forms of this organ were found among the nematodes from the Sevastopol bays and the Donuzlav Lake. There is a need for interdisciplinary research of this phenomenon in the natural and anthropogenic zones with extreme conditions in the Black Sea and other water basins for the understanding of the origin of this process and the ability of retention the acquired signs on the way of microevolution.

At present, we might only assume that we observe the process of either atavism or modern morpho-physiological adaptations to the changing environment. Obviously, you can assume that the female and male specimens having amphids with two elements of various forms can produce next - generations at three different branches with the set of elements: a) with one pair of species-specific form amphid; b) with one pair not specific form for this species and c) with different elements the same as the parent specimen.

The issues indicated in this chapter require future research:

1. What is the specific cause of the amphid abnormalities in nematodes?
2. Are these amphid abnormalities offer an advantage or a disadvantage compared to the "normal" organs for organisms inhabiting polluted ecosystems or a life under the impact of climate change?
3. Why similar morphological abnormalities are not recorded in nematodes from the other seas? Are these chemoreceptor organ abnormalities in nematodes unique for the Black Sea?
4. Is this phenomenon shows the nascent process of microevolution of nematode fauna in the Black Sea?

ACKNOWLEDGMENTS

This work was funded by many programs of the National Academy Science of Ukraine and the Academy Science of Russia. A part of this work was supported from EC 6th FP project HERMES GOCE-CT-2005-511324, and 7th FP project HYPOX 226213 and a joint project (108Y340) supported by

TUBITAK-NASU. We are grateful to Professor Antje Boetius for providing an opportunity to participate in RV 'Meteor' cruise 72/2 and RV 'M.S. Merian' cruise 15/1 and to Professor M.N. Cağatay for providing an opportunity to participate in RV 'Arar' cruise. We thank to Professors V. Egorov and S. Gulin for providing an opportunity to participate in cruises RV 'Professor Vodyanitsky'. We sincerely thank all colleagues of IBSS NASU and IBMR RAS for their cooperation in the material collecting. Special thanks to our colleagues in the Benthos Ecology Department of IBSS and IBMR. Lead engineer L. Lukyanova is acknowledged for her help in processing bottom sediment samples. We are grateful to anonymous reviewers for comments that helped to improve the manuscript.

REFERENCES

[1] Vinogradov, M. E., Shushkina E. A., Flint M. V. and Tumantseva, N. I. (1986). Distribution of plankton in the lower part of oxygen zone in the Black Sea. *Oceanology*, *26,* 300–309. (In Russian).

[2] Tchesunov, A. V. (2001). Life of marine Nematodes. *Soros educational journal. Biology*, *7 (6)*, 17–25. (*In Russian*).

[3] Filipjev, N. I. (1922). New data on free-living nematodes of the Black Sea. *Transactions of Stavropol Agricultural Institute, 1(16),* 83–184. (In Russian).

[4] Belogurov, O. I. (1982). Organization, Evolution, Systematics Oncholaimins (Nematoda: Enoplida: Oncholaimina) and some general questions about the morphology of nematodes: Autoabstr. dis... doct. biol. nauk., Vladivostok. (In Russian).

[5] Belogurov, O. I. and Belogurova, L. S. (1979). About morphological abnormalities (ugliness) in Nematodes and their possible evolutionary significance. *Journal of Zoology*, *58(II),* 1730–1733. (In Russian).

[6] Zahidov, M. T. (1973). Cases of abnormalities anomalies in free-living Nematodes in the Curonian Lagoon. *Journal of Zoology*, 52(10), 1567–1568. (In Russian).

[7] Mulvey, R. H. (1963). Abnormalities in the reproductive organs of a predaceous nematode *Prionchulus muscorum*. *Canadian Journal of Zoology*, *41(8),* 793–795.

[8] Timm, R. W. (1954). An abnormality of Oncholaimus marinus (Nematoda: Oncholaiminae). *Proceedings of the Helminthological Society of Washington,* 21(1), 36.

[9] Tsalolihin, S. Ya, (1975) A case of bivulvar Hermaphroditism in *Tobrilus bekmane* (Nematoda, Enoplida). *Journal of Zoology, 54(7),* 1084–1085. (In Russian)

[10] Galtsova, V. V. (1987). On the trends in the evolution of species and their enantsiomorphiya in free-living nematodes. *Journal of Zoology, 66(6),* 814–819. (In Russian).

[11] Gagarin, V. G. (1988a). Morphological variability *Eumonhystera filiformis* (Bastian, 1865) and *E. dispar* (Bastian, 1865) (Nematoda: Monhysteridae). *Ecology and morphology of aquatic invertebrates.* IBIW AS USSR, Borok, Dep. VINITI, 26.09.88 № 7151–B88, 204–213. (In Russian).

[12] Gagarin, V. G. (1988b). Morphological variability of free-living nematodes of the family Tobrilidae (Nematoda: Enoplida). *Ecology and Morphology of aquatic invertebrates.* IBIW AS USSR, Borok. Dep. VINITI, 27.09.88 № 7151–B88, 214–225. (In Russian).

[13] Daday, E. (1901). Microskopische Süsswassertiere aus Deutsch-Neu-Guinea. *Termesz. Fuzetck., 24,* 1–56.

[14] Tsalolihin, S. Ya. (1977). Nematoda with four amphids. *In:* E. S. Kirjanova, S. Ja. Tsalolihin (Eds.), *Free-living soil nematode, entomophilous and phytonematodes.* Leningrad, 108. (In Russian).

[15] Nicholas, W. L. (1984). The *biology of free-living nematodes.* Oxford Clarendon Press., 1–251.

[16] Riemann, F. (1972). Corpus gelatum und ciliäre Structuren als lichtmikroskopisch sichtbare Bauelemente des Seitenorgans freilebender Nematoden. *Zeitschrift fuer Morphologie der Tier, 72,* 46–76.

[17] Zur Strassen, O. I. (1904). *Anthraconema,* eine Gattung freilebender Nematoden. *Zoologische Jahrbuecher, 7,* 301-346.

[18] Inglis, W. G. (1964). The structure of the nematode cuticle. *Proceedings of the Zoological Society of London, 143,* 465–502.

[19] Burr, A. N. and Burr, S. (1975). The amphid of the nematode *Oncholaimus vesicarius*: ultrastructural evidence for a dual function as chemoreceptor and photoreceptor. *Journal of Ultrastructure Research, 51(1),* 1–15.

[20] Galtsova, V. V. (1991). Meiobenthos in marine ecosystems (on the example of free-living nematodes). *Proceedings of the Zoological Institute ANUSSR, 224,* 1–240. (In Russian).

[21] Higgins, R. P. and Thiel, H. (1988). *Introduction to the Study of Meiofauna.* Smithsonian Institution Press: Washington D.C., London, 1–488.

[22] Sergeyeva, N. G. (2004). Structure and Distribution of Meiobenthos in the Region of Methane Gas Seeps from the Black sea bottom. *Hydrobiological Journal*, *40(6)*, 45–56.

[23] Sergeeva, N. G., Gooday, A., Mazlumyan, S. A., Kolesnikova, E. A., Lichtschlag, A., Kosheleva, T. N., Anikeeva, O. V. (2012). Meiobenthos of the oxic/anoxic interface in the south-western region of the Black Sea: abundance and taxonomic composition, In: A.V. Altebach, J.H. Bernhard, J. Seckbach (Eds.), *Anoxia: Paleontological Strategies and Evidence for Eukaryote Survival*, New York, Springer, 369–401.

[24] Sergeeva, N. G., Mazlumyan, S. A., Çăgatay, N., and Lichtschlag, A. (2013). Hypoxic Meiobenthic Communities of the Istanbul Strait's (Bosporus) Outlet Area of the Black Sea. *Turkish Journal of Fisheries and Aquatic Sciences*, *13*, 33–41.

[25] Ürkmez, D., Brennan, M L., Sezgin, M. and Bat, L. (2015). A brief look at the free-living Nematoda of the oxic/anoxic interface with a new genus record (*Trefusia*) for the Black Sea. *Oceanological and Hydrobiological studies*, 44(4), 539–551.

[26] Ürkmez, D., Sezgin, M., Ersoy Karaçuha, M., Öksüz, İ., Katağan, T., Bat, L., Dağlı, E., Şahin, F. (2016). Within-year spatio-temporal variation in meiofaunal abundance and community structure, Sinop Bay, the Southern Black Sea. *Oceanological and Hydrobiological Studies,* 45, 1, 55–65.

[27] WoRMS Editorial Board (2016). World Register of Marine Species. Available from (http://www.marinespecies.org) at VLIZ. Accessed 2016-03-10.

[28] Egorov, V., Luth, U., Luth, C. and Gulin, M. B. (1998). Gas seeps in the submarine Dnieper Canyon, Black Sea: acoustic, video and trawl data. In U. Luth, C. Luth, H. Thiel (Eds.), *MEGASEEPS Gas Explorations in the Black Sea*, Project Report. Zentrum für Meres- und Klimaforschung der Universität Hamburg, Hamburg, 11–21.

[29] Egorov V.N., Polikarpov G.G., Gulin S.B., Artemov Yu. G., Stokozov N.A. and Kostova S.K. (2003) Present-day views on the environment-forming and ecological role of the Black Sea methane gas seeps. *Marine Ecological Journal*, 3, 5–26 (*in Russian*).

[30] Sergeeva N.G. Morphological variability of *Terschelling longicaudata* de Man, 1907 (Nematoda, Monhysterida) in the Black Sea/In: Fauna, Morphology and Systematics of lower Worms. *Proceedings of the Institute of Biology of Inland Waters*. 1991, 64(67), 117–129.

[31] Sergeeva N.G. 1992 Anatomical and morphological changes of polychaetes and nematodes/*In: Long-term changes in zoobenthos of the Black Sea*/Ed. VE. Zaika, Kiev: Naukova Dumka, 14-17 [in Russian].

[32] Sergeeva, N. G. (2003a). Morphological anomalies of nematodes as indices of the state environmental. *In:* V. N. Eremeev, A. V. Gaevskaya (Eds.), *Modern condition of biological diversity in near-shore zone of Crimea (the Black Sea sector))*, Sevastopol. Ekosi-Gidrophizika, 302-309. (In Russian).

[33] Hornafius, J. S., D. C. Quigley, and B. P. Luyendyk (1999), The world's most spectacular marine hydrocarbon seeps (Coal Oil Point, Santa Barbara Channel, California): Quantification of emission, *J. Geophys. Res., 104(C9),* 20,703 – 20,711.

[34] Polikarpov, G. G., Egorov, V. N., Nezhdanonv, A. I., Gulin, S. B., Kulev, Y. D. and Gulin, M. B., (1989). Methane gas seeps in the Black Sea — a new object of molismology. In G.G. Polikarpov (Ed.), *Molismology of the Black Sea*. Nauka, Kiev, 10–20. (In Russian).

[35] Luth, C., Luth, U., Gebruk, A. V. and Thiel, H. (1999). Methane gas seeps along the oxic/anoxic gradient in the Black Sea: Manifestations, biogenic sediment compounds, and preliminary results on benthic ecology. *Marine Ecolology, 20,* 221–249.

[36] Thiel, H., Schriever, G., Ahnert, A., Bluhm, H., Borowski, C., Vopel, K. (2001). The large-scale environmental impact experiment DISCOL – reflection and foresight. *Deep-Sea Research Part II, 48*, 3869–3882.

[37] Peckmann, J., Reimer, A., Luth, U., Luth, C., Hansen, B. T., Heinicke, C., Hoefs, J. and Reitner, J. (2001). Methane-derived carbonates and authigenic pyrite from the northwestern Black Sea. *Marine Geology, 177*, 129–150.

[38] Michaelis, W., Seifert, R., Nauhaus, K., Treude, T., Thiel, V., Blumenberg, M., Knittel, K., Gieseke, A., Peterknecht, K., Pape, T., Boetius, A., Amann, R., Jørgensen, B. B., Widdel, F., Peckmann, J. R., Pimenov, N. V. and Gulin, M. B. (2002). Microbial reefs in the Black Sea fueled by anaerobic oxidation of methane. *Science*, *297*, 1013–1015.

[39] Luth, U., Luth, C. (1998). Benthic meiofauna and macrofauna of a methane seep area south-west of the Crimean Peninsula, Black Sea. In U. Luth, C. Luth, H. Thiel (Eds.), *MEGASEEPS-methane gas seeps exploration in the Black Se*a, *14*. Berichte aus dem Zentrum für Meeres- und Klimato forsch, Hamburg, 113–126.

[40] Sergeeva, N. G. (2003b). Meiobenthos in the region with the methane gas seeps. In V. N. Eremeev and A. V. Gaevskaya (Eds.), *Modern*

condition of biological diversity in near-shore zone of Crimea (the Black Sea sector)), Sevastopol, Ekosi-Gidrophizika., 258–267. (*In Russian*).

[41] Sergeeva, N. G. (2004). Structure and Distribution of Meiobenthos in the Region of Methane Gas Seeps from the Black Sea Bottom. *Hydrobiological Journal*, *40(6)*, 45-56.

[42] Sergeeva, N. G., Gulin, M. B. (2007). Meiobenthos from an active methane seepage area in the NW Black Sea. *Marine Ecology*, *28*, 152-159.

[43] Lichtschlag, A., Janssen, F., Wenzhofer, F., Holtappels, M., Struck, U., Jessen, G. and Boetius, A. (2012). Effect of bottom water oxygenation on oxygen consumption and benthic biogeochemical processes at the Crimean Shelf (Black Sea). *Geophysical Research Abstracts*, *14*, EGU2012-9856.

[44] Lichtschlag, A., Donis, D., Janssen, F., Jessen, G.L., Holtappels, M., Wenzhöfer, F., Mazlumyan, S., Sergeeva, N., Waldmann, C. and Boetius, A. (2015). Effects of fluctuating hypoxia on benthic oxygen consumption in the Black Sea (Crimean shelf). *Biogeosciences, Supplement of Biogeosciences*, *12*, 5075–5092.

[45] Faschuk, J. D. (1995). Hydrogen sulfide zone of the northwestern Black sea shelf; the nature, causes, mechanisms of dynamics. *Water resourses*, *22(5)*, 568–584. (In Russian).

[46] Zaitsev Yu. P., Garkavaya G. P., Makarov Yu. N. et al. (1985). The Ecosystem of the North-Western part of the Black Sea under conditions of anthropogenic impact. *Procc. of the 3rd USSR Symposium "Anthropogenic eutrophication of natural waters*," Chernogolovka AN USSR, 49–72.

[47] Sergeeva N. G., Mazlumyan S. A. (2006). Meiobenthos of the Northwestern Shelf of the Black Sea in the Vicinity of the Zernov Phyllophora Field. *Hydrobiological Journal*, 42(5), 3–11.

[48] Sergeeva, N. G, Kharkevych, Kh. O, Revkova, T. N. (2013). Modern Structure of Meiobenthos of north-western shelf of the Black Sea. *IGCP 610 First Plenary Meeting and Field Trip,* "From the Caspian to Mediterranean: Environmental Change and Human Response during the Quaternary" (2013 - 2017), Tbilisi, Georgia, 12-19 October, 126–129.

[49] Emelyanov, V. A. (2013). Conclusion. In A. Y. Mitropolsky (Eds.), Geological, geo-environmental, hydro-acoustic and hydroecological investigations of the shelf and continental slope of the Ukrainian sector of the Black Sea. Kiev, 135-139. (In Russian).

[50] Revkov, N. K., Boltachova, N. A. and Alyomov, S. V. (2013). Macrozoobenthos. In A. Y. Mitropolsky (Eds.), *Geological, geo-environmental, hydro-acoustic and hydroecological investigations of the shelf and continental slope of the Ukrainian sector of the Black Sea.* Kiev, 124-134. (In Russian).
[51] Sergeeva N. G. (2013). Meiobenthos. In A. Y. Mitropolsky (Eds.), *Geological, geo-environmental, hydro-acoustic and hydroecological investigations of the shelf and continental slope of the Ukrainian sector of the Black Sea.*, Kiev, 119-124.
[52] Burgess, R. M., Terletskaya, A. V., Milyukin, M. V., Povolotskii, M., Demchenko, V. Y., Bogoslavskaya, T. A., Topkin, Y. V., Vorobyova, T. V., Petrov, A. N., Lyashenko, A., Ho, K. T. (2009). Concentration and distribution of hydrophobic organic contaminants and metals in the estuaries of Ukraine. *Marine Pollution Bulletin, 58*, 1103–1115.
[53] Burgess, R. M., Konovets, I. M., Kipnis, L. S., Lyashenko, A. V., Grintsov, V. A., Petrov, A. N., Terletskaya, A. V., Milyukin, M. V., Povolotskii, M, Demchenko, V. Y., Bogoslavskaya, T. A., Topkin, Y. V., Vorobyova, T. V, Portis, L. M., Ho, K. T. (2011). Distribution, magnitude and characterization of the toxicity of Ukrainian estuarine sediments. *Marine Pollution Bulletin, 62*, 2442–2462.
[54] Zaika, V. E., Konovalov, S. K., Sergeeva, N. G. (2011). The events of local and seasonal hypoxia at the bottom of Sevastopol bays and their influence on macrobenthos. *Marine Ecological Journal, 10(3)*, 15-25. (*In Russian, English summary*).
[55] Ivanov, V. A., Ovsyanyi, E. I., Repetin, L. N., Romanov, A. S. and Ignatieva, O. G. (2006). *Hydrologo-hydrochemical regime of the Sevastopol Bay and its climatic and anthropogenic changes, Sevastopol*, EKOSI-Hydrophysics, 1–90. (In Russian).
[56] Orekhova, N. A. and Konovalov, S. K. (2009). Oxygen and hydrogen sulphide in the bottom sediments Sevastopol Bay. *Proceedings of Marine Hydrophysical Institute*, Sevastopol, 48–56. (In Russian).
[57] Ignatieva, O. G., Ovsyanyi, E. I. and Romanov, A. (2002). Studies the chemical composition and Assessment of the level of pollution of bottom sediments in the Sevastopol Bay (the Results of the study 2001–2002). *Sustainable development of territories of South Russia and Ukraine: Evolution, Functioning and Resources/Collection of scientific works of MSU. M. V. Lomonosov of the Black sea branch.* Sevastopol, SPC "ECO-Hydrophysics," 103–125 (In Russian).

[58] Sergeeva, N. G., Mazlumyan, S. A. and Konovalov, S. K. (2010). Benthic fauna depending on oxygen depletion (inner and outer parts of Sevastopol Bay). In V. Yanko-Hombach (Eds), *Extended Abstracts of the* Sixth *Plenary Meeting and Field Trip of IGCP-521 "Black Sea - Mediterranean corridor during the last 30 ky: Sea level change and human adaptation" - INQUA 0501 "Caspian-Black Sea-Mediterranean Corridor during last 30 ky: Sea level change and human adaptive strategies.*" Rhodes, Greece, 27 September - 5 October, 187–189.

[59] Sergeeva, N. G., Kolesnikova, E. A. (2014). Meiobenthos structure under different anthropogenic conditions at the coastal zone of SW Crimea (Sevastopol Bays, Black Sea). In V. Yanko-Hombach (Eds.), *Proc. IGCP 610 "From the Caspian to Mediterranean: Environmental Change and Human Response during the Quaternary" IGCP 610 Second Plenary Meeting and Field Trip,* Baku, Azerbaijan, 12-20 October, 120–123.

[60] Zaika, V. E., Ivanova, E. A., Sergeeva, N. G. (2011). Seasonal changes of meiobenthos of the Sevastopol bays with the analysis of influence of bottom hypoxia. *Marine Ecological Journal*, Special issue, *2*, 29–36. (In Russian, English summary).

[61] Sergeeva, N. G., Kolesnikova, E. A., Latushkin, A. A., Chepyzhenko, A. A. (2014). Diversity of meiobenthos in the soft ground of the Sivash Lake. In: Proceedings of the 3rd Intern. Scientific-Practical Conf. "Biodiversity and Sustainable Development" Simferopol, on May 26 - June 1, Simferopol, 84–85.

[62] Özsoy, E. and Ünlüata, Ü., (1997). Oceanography of the Black Sea: a review of some recent results. *Earth-Science Reviews, 42*, 231–272.

[63] Özsoy, E., Ünlüata, Ü. and Top, Z. (1993). The Mediterranean water evolution, material transport by double diffusive intrusions, and Interior Mixing in the Black Sea. *Progress in Oceanography*, *31*, 275–320.

[64] Sergeeva, N. G., Mazlumyan, S. A., Çăgatay, N., and Lichtschlag, A. (2013). Hypoxic Meiobenthic Communities of the Istanbul Strait's (Bosporus) Outlet Area of the Black Sea. *Turkish Journal of Fisheries and Aquatic Sciences*, *13*, 33–41.

[65] Tsitsugina W. G., Gorbenko W. P., Zherko N. W., Svetasheva S. K. (1988). Natural chromosomal mutagenesis of the contour bionts of the open waters of the Black Sea in modern conditions. Abstracts 3 All-Union Conf. Marine Biology, Kiev, 192.

[66] Losovskaya, G. V. (1988). Long-term changes in the composition and distribution of polychaetes of the North-Western Black Sea. *Hydrobiologycal Journal, 24(4)*, 21-25.

[67] Gourbault N. and Renaud-Mornant J. (1990). Micro-Meiofaunal Community Structure and Nematode Diversity in a Lagoonal Ecosystem (Fangataufa, Eastern Tuamotu Archipelago). *Marine Ecology, 11(2)*, 175-189.

[68] Burton, G. A. and Scott, Jr. K. J. (1992). Sediment Toxicity evaluation their niche in ecological assessments. *Environmental Science and Technology Environ, 26 (11)*, 2068–2075.

[69] Adams, W. J., Kimerle, R. A. and Barnett, J.W. (1992). Direct transfer of chemicals from sediments to organisms is now considered a major route of exposure for many species. *Environmental Science and Technology Environ, 26 (10)*, 1865–1875.

[70] Klerks P. I., Weis J. S. (1987). Genetic Adaptation to Heavy Metals in Aquatic Organisms: A Review. *Environ. Pollut.*, *45*, 173-205.

[71] Sergeeva NG and Mazlumyan SA (1988) The structure of the benthic communities in conditions chronic pollution IBSS NANU, Sevastopol, Dep. VINITI, 1988, 29.07.88, № 6569–B88, 1-15.

[72] Fadeev, V. I. and Fadeeva, N. P. (1999). Distribution of small-sized benthic organisms in conditions of chronic oil pollution of bottom sediments. *Proceed. Intern. Simp.Earth-Water- Humans.* Kanazawa, 146–154.

[73] Fadeeva, N. P., Bezverbnaya, I. P., Tasaki, K., Vatanabe, N. and Fadeev, V. I. (2002). Composition, Structure and Metabolism of benthic Communities silty gound under conditions of chronic anthropogenic pollution (for example, Bay Golden Horn). Fundamental research of marine biota. Biology, Chemistry and Biotechnology. Proc. Conf. of students, graduate students and young scientists of REC FENU"Marine Biota. Vladivostok: Publishing house of Far Eastern University. University Press, 62–65. (In Russian).

[74] Tietjen, J. H. (1977). Population distribution and structure of the free-living nematodes of Long Island Sound. *Marine Biology, 43*, 123-136.

[75] Heip, C., Vincx, M. and Vranken, G. (1985). The ecology of marine nematodes. *Oceanography and Marine Biology*: *An Annual Review*, *23*, 399–489.

[76] Vranken, G., Heip, C. (1986). The productivity of marine nematodes. *Ophelia*, *26*, 429–442.

[77] Nasira, K., Shahina, F., Kamran, M. (2010). Response of free-living marine nematode community to heavy metal contamination along the coastal areas of Sindh and Balochistan, Pakistan. *Pakistan Journal of Nematol Olgy, 28*, 263–278.

[78] Thiermann F., Vismann B. and Giere O. (2000). Sulphide tolerance of the marine nematode Oncholaimus campylocercoides – a result of internal sulphur formation? *Marine Ecology Progress Series, 193*, 251-259.

[79] Sergeeva N. G., Mazlumyan S. A., Lichtschlag, A. and Holtappels, M. (2014). Benthic Protozoa and Metazoa Living under Anoxic and Sulfide Conditions in the Black Sea: Direct Observations of Actively Moving Ciliophora and Nematoda. *International Journal of Marine Science* 2014, 4(42), 1-11. http://ijms.biopublisher.ca.

[80] Sergeeva N. G. and Mazlumyan S. A. (2015). Deep - water hypoxic meiobenthic Protozoa and Metazoa taxa of the Istanbul Strait's (Bosporus) outlet area of the Black Sea. *Ecologica Montenegrin.*, 2 (3), 255-270.

[81] Burdin, K. S. (1985). *Basics of Biological Monitoring. Moscow*, MSU, 158. (In Russian).

[82] Malakhov V. V. (1986) Study of the structure of the sense organs of free-living marine nematodes. 2. The senses and the evolution of sensory structures in nematodes. *Journal of Zoology, 62(2*), 191-200.

[83] Malakhov V. V. (1986). *Nematodes: Structure, Development, System and Phylogeny*. Moscow, Nauka, 215.

[84] Davydkova, I. I. and Fadeeva, N. P. (2003). Peculiarities of the life-cycle of the free-living marine nematode *Oncholaimium ramosum* (Enoplida: Oncholaimidae) from the Sea of Japan. *Russian Journal of Nematology, 11(21)*, 135. (In Russian).

[85] Alekperov, U. K. (1989). Antimutagens and protection of the gene pool. Moscow, Znanie, 63. (In Russian).

[86] Schmidt S. (2009). Shall we really do it again? The powerful concept of replication is neglected in the social sciences. *Review General Psychology, 13*, 90–100.

[87] Nakagawa, Sh. and Parker, T. H. (2015). Replicating research in ecology and evolution: feasibility, incentives, and the cost-benefit conundrum. *BMC Biology, 13*, 88.

[88] Severtsov, A. N. (1967). The main directions of the evolutionary process. *Morphological evolution*. Moscow State University, 140–172.

[89] Zotin, A. I. and Zotin, A. A. (1999). The direction, rate and mechanisms of progressive evolution. *Thermodynamic and experimental bases*. Moscow, Nauka, 319.

[90] Severtsov, A. S. (1987). Criteria and conditions of aromorphic organization. In L. P. Tatarinov, A. P. Rasnitsyn (Eds.), *Evolution and biocenotic crises*. Moscow, Nauka, 64–76.

[91] Yablokov, A. V. and Yusufov, A. G. (1998). *The doctrine of evolution*. Moscow, Higher school, 31.

Reviewed by

Dr. Alexei V. Tchesunov, Professor, Dr. Sci.
Dept. Invertebrate Zoology, Biological Faculty, Moscow State University, Vorobyovy Gory, Moscow 119 991, Russia
fax: +7 495 9394495; mailto: AVTchesunov@yandex.ru

Reviewed by

Professor Marian-Traian Gomoiu
National Institute of Marine Geology and Geo-Ecology (GeoEcoMar), Constanta Branch, 304 Mamaia Blvd., 900581 Constanta, Romania
Marian-Traian Gomoiu <mtg@cier.ro>

Biographical Sketch

Name: Derya ÜRKMEZ

Affiliation:
Scientific and Technological Research and Application Center, Sinop University.

Education: PhD

Address: Scientific and Technological Research and Application Center, Sinop University, TR57000, Near the city airport, Sinop-TURKEY

Research and Professional Experience:

1- Training and Education:

Years from...to...	Name of establishment, Town, Country	Certificates, diplomas and/or grades obtained	Main subjects
2009-2015	Sinop University, Institute of Natural and Applied Sciences, Sinop-TURKEY	PhD	Free-living marine nematodes, Meiobenthos ecology
2000-2003	Ege University, Institute of Natural and Applied Sciences, İzmir-TURKEY	Master of Science	Marine benthos ecology (Macrobenthos)
1996-2000	Ege University, Faculty of Fisheries, Department of Marine Sciences, İzmir-TURKEY	Bachelor of science	Marine Sciences

2- International Projects Participated

1) Comparative studies of meiobenthic community structure off Crimea (Ukraine) and Sinop (Turkey) with respect to identification of Marine Protected Areas (MPAs) in Turkish Black Sea, 2009-2011, TÜBİTAK (Scientific and Technological Research Council of Turkey)-NASU (Ukranian Academy of National Sciences,) Joint Project, Project No: 108Y340.
2) CoastLearn – Black Sea (CLBS) (2010-2012). Integrated Coastal Area Management Education Project - Leonardo Da Vinci Life Learning Programme- project number 2010-1-TR1-LEO05-16745.
3) CoCoNET - FP7 Project—Towards COast to COast NETworks of marine protected areas (from the shore to the high and deep sea), coupled with sea-based wind energy potential. (2011-) Call identifier FP7-OCEAN-2011 Collaborative project; project number: 287844.

4) MISIS - EU ENV-BLACK SEA Project: MSFD (Marine Strategy Framework Directive) Guiding Improvements In the Black Sea Integrated Monitoring System- MISIS- Nr. 07.020400/2012/616044/ SUB/D2.
5) Meiofauna as an environmental bioindicator in marine ecosystems of Turkey and Montenegro – TÜBİTAK (Scientific and Technological Research Council of Turkey) –MoS (Ministry of Science of Montenegro), 2015- ongoing.

Professional Appointments (Work experience):

Dates (from... to...)	Employer	Position
2015- cont.	Sinop University – Scientific and Technological Research and Application Center	Specialist Doctor
2010-2015	Sinop University – Faculty of Fisheries – Department of Marine Biology and Ecology	Research Assistant
2004-2009	Freelance	Medical Interpreter (English-Turkish)
2000-2003	Ege University – Faculty of Fisheries Department of Marine Biology and Ecology	Research Assistant

Honors:

First Place among young scientists award –Sixth International Environmental “Micropaleontology, Microbiology and Meiobenthology” Congress, EMMM-2011, Moskow-RUSSIA.

BIODIVMEX Workshop Grant (CNRS-French National Center for Scientific Research)-Malta, 2012.

Wageningen University bursary to attend 31st International ESN Symposium in Adana, Turkey– European Society of Nematologists, (Nematology group-Netherlands)- 2012.

Publications Last 3 Years:

Ürkmez D. 2015. Dynamics Of The Meiobentic Nematode Fauna Of Sinop Bay And The Taxonomy And Ecology Of Chromadoridae, PhD thesis, 342p, Sinop University, Sinop-TURKEY.

Ürkmez D, Sezgin M. 2016. *Micoletzkyia longispicula* Huang and Cheng, 2011 (Nematoda: Phanodermatidae) in Turkey: First record outside the type locality. Ecologica Montenegrina. 6: 66-69. http://www.biotaxa.org/em/article/view/20113/19504.

Ürkmez D, Sezgin M, Ersoy Karaçuha M, Öksüz İ, Katağan T, Bat L, Dağlı E, Şahin F. 2016. Within-year spatio-temporal variation in meiofaunal abundance and community structure, Sinop Bay, Southern Black Sea. Oceanological and Hydrobiological Studies. 45 (1): 55-65. DOI: **10.1515/ohs-2015-0006.**

Ürkmez D, Brennan M, Sezgin M, Bat L. 2015. A brief look at the Free-living Nematoda of the oxic/anoxic interface with a new genus record (*Trefusia*) to the Black Sea. Oceanological and Hydrobiological Studies. 44(4):539-551 · DECEMBER 2015. DOI: 10.1515/ohs-2015-0051.

Vorobyova L, **Ürkmez D,** Sezgin M. 2015. Meiobenthos Manual. Biological Methodology in the Black Sea. Technical Report of the EU MISIS project (MSFD Guiding Improvements in the Black Sea Integrated Monitoring System).

Ürkmez D, Sezgin M, Bat L. 2014. Use of nematode maturity index for the determination of ecological quality status: a case study from the Black Sea. 2014. Journal of Black Sea/Mediterranean Science, Vol.20, No.2: 96-107.

Ürkmez D, Brennan ML. 2013. A new species of *Halaphanolaimus* (Nematoda: Leptolaimidae) from the Southern Black Sea (Turkey) with a modified key for species identification. Zootaxa, 3691 (2): 220-228. http://dx.doi.org/10.11646/zootaxa.3691.2.2.0.

Ürkmez D, Sezgin M, Bat L. 2013. Attempt for the use of nematode diversity to evaluate the ecological quality status: an example for the Black Sea. 4th Black Sea Scientific Conference, BSGES-2013, “Challenges Towards Good Environmenal Status,” Constanta-Romanya. S. 114.

Sezgin M, **Ürkmez D**, Ersoy Karaçuha M. 2013. Application of Nematode/Copepod Index in the Black Sea: Any indication for use in monitoring of organic pollution? 4th Black Sea Scientific Conference, BSGES-2013, “Challenges Towards Good Environmenal Status,” Constanta-Romanya. S. 41.

Raineault N, Ramey-Balci P, Shank TM, Bors E, Duman M, **Ürkmez D**, Tüzün S. 2013. Exploration of the Anaximander Mountains: Mud Volcanoes, Cold-Seep Communities, and Cold water Corals. Oceanography (supplement), 26 (1), 30-35.

Brennan ML, Davis D, Roman C, Buynevich I, Catsambis A, Kofahl M, **Ürkmez D**, Vaughn I, Merrigan M, Duman M. 2013. Ocean dynamics and anthropogenic impacts along the southern Black Sea shelf examined through the preservation of pre-modern shipwrecks. Continental Shelf Research, 53: 89-101.

Pešić V, Sezgin M, Ersoy Karaçuha M and **Ürkmez D**. 2013. New records of marine water mites (Acari: Hydrachnidia, Pontarachnidae) from the southern Black Sea (Sinop Bay, Turkey). Mediterranean Marine Science, 14 (1), 45-47.

In: Baltic and Black Sea
Editor: Andre Mills
ISBN: 978-1-63485-233-3

Chapter 4

FISH BIODIVERSITY: THE ROLE OF DEFENSE SYSTEMS TO THE ADAPTATION TO ECOLOGICAL CONDITIONS IN THE BLACK SEA ECOSYSTEM

I. I. Rudneva[1,*], E. N. Skuratovskaya[1], I. I. Chesnokova, T. B. Kovyrshina[1], S. O. Omelchenko[2] and I. N. Zalevskaya[3]

[1]The A.O. Kovalevsky Institute of Marine Biological Research of RAS, Sevasopol, Russia
[2]The Crimean Regional in – service Teacher Training Institute Republic of Crimea, Simferopol, Russia
[3]Vernadsky Crimean Federal University, Simpheropol, Russia

ABSTRACT

Black Sea is one of the largest indoor anoxic sea of the world. The specific ecological conditions of the Black Sea, including anoxic water containing high concentrations of the hydrogen sulfide, play an important role in fish adaptations to unfavorable natural factors and need to study the adaptive mechanisms of their inhabiting in different biotops. In addition, great pollution level in the Black Sea caused high anthropogenic

---

* Corresponding Author address: svg-41@mail.ru.

activity, especially in the coastal area, influences directly and indirectly on fish populations and their health. Assessment of fish natural populations health is a great problem because normal values range widely and depend to natural ecological conditions and their fluctuations. Biomarker analysis of field-collected species can provide information on their physiological status and environment. Therefore, the main idea of the chapter is to study the role of blood biomarkers of highly distributed Black Sea fish species belonging to different ecological groups to the different natural and anthropogenic impacts on marine environment, to determine their main trends and importance for fish adaptations for keeping their life with the changes of ecological conditions. For this purpose the biochemical parameters of various highly distributed Black Sea fish species inhabiting different biotops were studied and the responses of biomarkers were detected. The levels of heavy metals in examined fish species tissues were determined and the responses of blood biomarkers were observed. The biomarkers studies help in understanding the relationship of blood biochemical characteristics to the habitat and adaptability of the fish species to the environment. The principles and perspectives of the development of permanent monitoring programs in the Black Sea for fish biodiversity protection used biomarkers are discussed.

Keywords: fish, biomarkers, adaptation, biotops, oxidative stress, trace elements, monitoring

1. Introduction

Marine resources play an important role for human existence. Anthropogenic activity impacts marine communities and entire ecosystem. Pollution of marine water and sediments deficiency of oxygen, high contents of hydrogen sulfide and xenobiotics decrease fish and invertebrates biodiversity, change the ecological conditions of spawning areas and has resulted to mass death of bottom fish species, crustacea, mollusks and algae. Black Sea communities were transformed very strongly caused high anthropogenic activity in its basin. Alterations in marine communities were demonstrated in phyto- and zooplankton, macrophytes, ichthyoplankton, fish and benthic organisms (Kovalev, Gubanova, 1995; Rudneva, Petzold-Bradley, 2001; Rudneva, 2011).

Fish are very sensitive to anthropogenic pollution and some of them may be tested as biomonitors for the evaluation of the ecological status of marine environmental quality. During the period of 1950-1990s Black Sea ichthyophauna was changed and fish biodiversity declined, especially in coastal waters which are the main recipients of sewage. According the data of 1990-s, 143 fish species inhabited Black Sea. For instance, in Sevastopol bays fish species biodiversity was reduced more than 2- fold and their storage was declined more than 100-fold compared to the 1950s (Konovalov, 1996). Besides that, damage of maturation, spawning process, formation of bisexual gonads in several fish species (horse mackerel) demonstrate a negative transformation in Black Sea fish populations, resulted to decline of fish biodiversity and population size. However, at the beginning of 2000-s the biodiversity increased and 233 species were documented in Black Sea and at present (Boltachev and Karpova, 2012). Most of them inhabit coastal waters.

The problem is not only to detect catastrophic changes and negative consequences of unfavorable impact, but to continue the study all these effects in various levels of fish biological organization more carefully for the purpose of their evaluation, identification, and search the effective methods for restoration of the marine ecosystem. Therefore, for the evaluation of fish status and their habitats various biomarkers are used. Biomarker analysis of field-collected fish can provide information of their health and the status of the environment. The application of biomarkers is well-advanced for the marine environment in Western countries, whereas additional efforts are needed for the Black Sea environment and the riparian countries. However, criticisms that have mainly hampered a more general use of biomarkers are their lack of specificity and ecological relevance. It is important to cope with these problems because biomarkers have a special interest as a cost-effective, sensitive, and early warning system (Hansson et al., 2006; Amiard-Triquet and Berthet, 2015).

The resistance of aquatic organisms to contamination depends on many factors including their phylogenic position, ecological and biological characteristics, physiological status (Hotard and Zou, 2008) and the presence of efficient detoxification mechanisms. Previously we described the variations in antioxidant system in different tissues of some Black Sea fish species which reflect adaptive strategy of fish species and their ability to cope with the environment (Rudneva, 2012). The further study of the molecular mechanisms of fish resistance to environmental stress led the anthropogenic pollution is very important for the understanding the different ways of adaptations of aquatic organisms and for risk assessment.

Physical and chemical changes in the marine environment cause various physiological and biochemical alterations in fish and their health. A number of environment pollutants can cause oxidative stress in aquatic organisms. For instance, trace elements are concentrated in to fish in the marine water and food nets. Heavy metals pollution causes oxidative stress in fish because among them there are some transition metals like iron, copper, zinc, that through the Fenton reaction are able to facilitate the conversion of superoxide anion to hydroxyl radical which damage cell membrane and biomolecules (Livingstone, 2001; Martinez-Alvarez et al., 2005; Farombi et al., 2007; Lane, 2011). Other metals such as cadmium, lead and mercury may interact with SH-groups of the proteins and peptides including glutathione, blocked them and produce oxidative stress indirectly decreasing the level of glutathione or via metal-induced displacement of redox metal ions (Barata et al., 2005). Fish living in marine locations contaminated with different xenobiotics may be exposed an oxidative stressors caused be a variety of oxiradicals. To protect to this chemical compounds aquatic organisms have developed different mechanisms such as the induction of antioxidant enzymes including superoxide dismutase, catalase, peroxidase and glutathione (GSH) related enzymes (glutathione reductase (GR), glutathione peroxidase (GP) and glutathione-S-transferase (GST). GST is a phase II biotransformation enzyme which plays an important role in the removal of xenobiotics. It conjugates harmful electrophilic compounds with endogenous reduced glutathione to protect the nucleophilic molecules such as proteins and nucleic acids against oxidative damage (Di Giulio et al., 1989; Winston and Di Giulio, 1991). Alterations in the antioxidant enzyme activities of aquatic organisms in response to pollutants are used to indicate the potential for more severe hazards.

Alanine aminotransferase (ALT) and aspartate aminotransferase (AST) catalyse the interconversion of amino acids and α-ketoacids by transfer of amino groups. ALT catalyses the transfer of the amino group from alanine to α-ketoglutarate to form glutamate and pyruvate. AST catalyses the transfer of the amino group from aspartate to α-ketoglutarate to form glutamate and oxaloacetate (van der Oost et al., 2003). Both aminotransferases play an important role as a link between carbohydrate and protein metabolism. They are detected in blood serum and in cells in various tissues. An increase of enzyme activity in the blood serum, plasma and other extracellular fluids was observed in many investigations of the organisms impacted unfavorable conditions, and it is related to organ dysfunction or internal lesions in tissues. The damaged cells release their contents (including aminotransferases) towards the blood stream, and the level of these enzymes enhances in serum (Martinez-Porchas et al., 2011). AST is used as clinic diagnostic tool, and it is associated with cell necrosis of the liver and skeletal or cardiac muscle, starvation and lacking of vitamin E. Plasma ALT is an acute hepatic damage good marker (Coppo et al., 2001-2002). Along with the study of transaminase activity, several researchers used the method of multivariate approximation. One such approximation is the ratio proposed by de Rytis coefficient (AST/ALT). This index allows us to estimate the functional load on the liver and heart. Sudden changes in the coefficient of relative discriminatory value as indicators of pathological processes occurring in these organs, which cause displacement synthesis, release, and metabolic conversion of aminotransferases. Decrease of this value is indicative of massive hepatocytes, its increase is associated with myocardial infarction (Titov and Bochkova, 1990).

Cholinesterase (ChE) is an enzyme that was been widely used as a biomarker for neurotoxicity. It might be considered as a particularly sensitive marker for organophosphate (OP) and carbamate (CB) pesticides, but may respond to heavy metals, detergents and some algal toxins also (van der Oost et al., 2003; Assis et al., 2011).

Oligopeptides are short sequence peptides ranging from 2 to 20 amino acids. This class of proteins includes potent biologically active compounds, which can modulate various cellular and molecular processes. Among them the biological mediators are identified which play an important role in affecting of biological activity. Currently, the use of protein peptide fragments is a rapidly growing field of research. Studies suggest that various biologically active peptides can result in favorable or negative outcomes on organism, and the role of biologically active short sequence peptides as potential agents through the modulation of age- and sex-dependent biochemical characteristics. Specifically, oligopeptides are short sequence peptides including from 2 to 20 amino acids. Among them the biological mediators are identified which play an important role in affecting of biological activity (Grune, 2000). Level of oxidative modification of proteins in tissues are good biomarkers for evaluation of oxidative stress in fish caused anthropogenic impact (Luschak et al., 2005; Kovyrshina and Rudneva, 2012; Rudneva et al., 2012a).

Thus, various biochemical parameters are used for the evaluation of fish health, which reflect fish physiological status and both biotic and abiotic factors (including anthropogenic) cause their fluctuations (Martinez-Alvarez et al., 2005; van der Oost et al., 2003). Among them blood biochemical characteristics are used widely as early warning indicators of organism physiological and health status (Haman et al., 2012). The biomarkers studies help in understanding the relationship of blood biochemical characteristics to the habitat and adaptability of the fish species to the environment (Fazio et al., 2012; 2013a, b, c).

The aim of the present study is to detect interspecific differences in blood biomarkers in Black Sea high distributed fish species and their relation to the ecological status and exposure to environmental pollution. Therefore, in our study we try

- to document the baseline concentrations of six trace elements (Zn, Cu, Pb, Cd, Hg and As) in six teleost fish species caught in Sevastopol coastal area (Black Sea),
- to determine the relationships between trace elements concentrations and blood biomarkers;
- to show correlations between trace elements concentration, tested biomarkers and fish ecological status.

2. Materials and Methods

2.1. Materials

The following fish species were used: horse mackerel *Trachurus mediterraneus* (Staindachner) (Carangidae) (n = 38), high body pickarel *Spicara flexuosa* (Rafinesque) (Centracanthidae) (n = 35), red mullet *Mullus barbatus ponticus* (Essipov) (Mullidae) (n = 25), round goby *Neogobius melanostomus* (Pallas) (Gobiidae) (n = 62), toad goby *Mesogobius batrachocephalus* (Pallas) (Gobiidae) (n = 23), scorpion fish *Scorpaena porcus* (L.) (Scorpaenidae) (n= 40). The fish were carried out in Sevastopol coastal waters (Black Sea) during spring-autumn period 2005-20013 (Figure 1). The animals were transported to the laboratory in the containers with marine water and constant aeration.

For further investigations fish were grouped in four ecological groups according their belonging to different depth habitats and their biological, ecological and physiological peculiarities: benthic (scorpion fish, round goby, toad goby), suprabenthic (red mullet), suprabenthic/pelagic (high body pickerel) and pelagic (horse mackerel).

Sample Preparation

The blood was taken by caudal arteria puncture and serum was separated. The red blood cells (RBC) were processed as we described previously (Rudneva, 1997). The sediment was washed three times with 0.85% NaCl solution and then lyzed by addition of 5 vol of distilled water for 24 h at the refrigerator. The enzyme activity was determined in the lysates immediately after preparation. Fish fillets were frozen for further trace elements analysis.

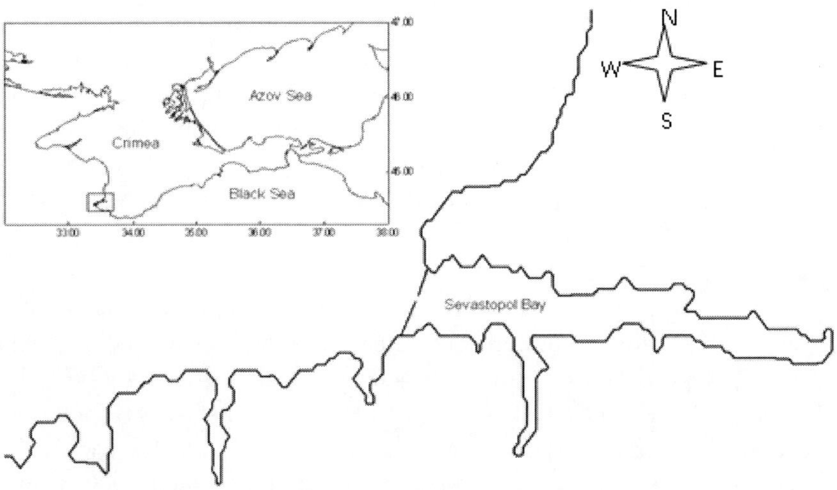

Figure 1. Sampling sites in Sevastopol Bay (Sevastopol, Black Sea).

2.2. Chemical and Biochemical Assays

Trace Elements Determinations

Metal concentration levels were determined in single animal fillet according the method described in Tuteliayn (1986). Fish tissues were dried for 9 hours at temperature increasing from + 50ºC to +450ºC. Dried samples were digested with concentrated nitric acid. Concentration of Cd, Cu, Pb and Zn was measured by atomic absorption method used Hitachi (Japan) spectrophotometer. Arsenic concentration was detected by spectrophotometric method using Ag-diethylditiokarbamat reagent in chloroform and calculated on the basis of calibration standards at the wave length of 520 nm.

Mercury concentration was determined on atomic absorption spectrophotometer "Julia-2" (Russia) (Tuteliayn, 1990). 5 g of the tissues was added in 1 ml ethanol, 5 ml distilled water, 5 ml concentrated nitric acid and incubated at room temperature 30 min. 5 ml concentrated sulphuric acid was added in the mixture and heated during 15-20 min at +70ºC, then 30 ml of bidistilled water was added and boiled on water bath during 15-20 min. 1 ml of the sample was added in the cuvette of the spectrophotometer and Hg concentrations was measured. All determinations were processed in triplicates.

Oligopeptide Concentrations (OP)

The concentration of oligopeptides (OP) was detected in blood serum of the tested fish species. Specifically, 0.25 ml Trichloracetic acid (TCA) was added to 0.5 ml of the sample and centrifuged at 9000g for 30 min. Next, 0.3 ml of the supernatant was mixed in 3.7 ml 3% NaOH and 0.2 ml Benedict reagent. The mixture was incubated for 15 min at room temperature and then optical density (OD) was measured at 330 nm (Karyakina and Belova, 2004). We express the results in arbitrary units (OD at 330 nm per mg protein x10^{-3}).

Oxidated Proteins Concentration

Oxidated proteins concentration in fish blood serum was determined spectrophotometrically according the method Dubinina et al., (1995) with some modifications. 0.05 ml serum was added 1 ml 20% TCA solution and 1 ml 2,4-dinitrophenilhidrazine (DNP) in 2 M HCl. The mixture was incubated 1hr at room temperature and then it was centrifuged at 3000 g due 15-20 min. Sediment was washed by ethanol-ethylacetat (1:1) solution and it was centrifuged again. 2.5 ml 8 M urea was added to dry sediment and it was solved in boiling water. Optical density (OD) of carbonyl groups was measured at 346 (neutral aldehyde), 370 (neutral ketone), 430 (basic aldehyde) and 530 nm (basic ketone) used spectrophotometer Specol-211 (Carl Zeiss, Iena, Germany). The results were shown in arbitrary units (OD per ml serum).

Antioxidant Activity (SOD, CAT, PER, GR) and GST)

Antioxidant activities in the red blood cells from the species in this study were determined according to methods described previously (Rudneva, 1997), with a few minor modifications.

Superoxide dismutase (SOD) was assayed on the basis of inhibition of the reduction of nitroblue tetrasolium (NBT) with NADH mediated by phenazine methosulfate (PMS) under basic conditions (Nishikimi et al., 1972). All measurements were performed in 0.017 M sodium pyrophosphate buffer pH 8.3 at +25ºC. The reaction mixtures contained 5 μM NBT, 78 μM NADH, 3.1 μM PMS, and a 0.1 ml sample; the final volume was 1.5 ml. The reaction was carried out in the cuvette of spectrophotometer at 560 nm.

Catalase (CAT) was measured by the method involving the reaction of hydroperoxide reduction.

Peroxidase (PER) activity was detected by spectrophotometric method using benzidine reagent (Litvin, 1981). The reaction mixture contained 1 ml acetate buffer pH 5,4, 0.4 ml 0.09% benzidine, 0.2 ml 0.03% H_2O_2, and 0.2 ml sample. The reaction followed in a spectrophotometer for 1 min at 20ºC and at 600 nm.

Glutathione reductase (GR) activity was assayed spectrophotometrically using a method modified after Goldberg and Sparner (1987). The reaction mixture contained 0.1 ml mM NADPH, 0.5 ml 7.5 mM oxidized glutathione, 0.2 ml mM EDTA, and 2 ml 0.05 M phosphate buffer pH 8.0. After incubation for 10 min, the extinction of the mixture was determined at 340 nm.

Glutathione-S-transferase (GST) activity was determined by the method of Habig et al., (1974) by following the increase in absorbance at 340 nm due to the formation of the conjugate 1-chloro-2,4-dinitrobenzene (CDNB) using as substrate at the presence of reduced glutathione (GSH). The reaction mixture was prepared by mixing 1.5 ml sodium phosphate buffer 0.1 M pH 6.5, 0.2 ml GSH 9.2 mM, 0.02 ml CDNB 0.1 M and 0.1 ml of the sample. The absorbance was measured at 340 nm and at the temperature +25^{0}C spectrophotometrically using Specol-211 (Germany). The increase in absorbance was recorded for a total 3 min. The reaction solution without the fish hepatic homogenates was used as blank.

Aminotransferases Activity

Aminotransferases (ALT and AST) activity was also determined spectrophotometrically with 2,4-dinitrophenylhydrazine using the standard kit (Filicit - Diagnosis, Ukraine). Specifically, 0.2 ml substrate-buffer solution was added to 0.04 ml of liver extract or serum and incubated at room temperature for 1 h. The reaction was stopped by 0.2 ml 1,4-dinitrophenylhydrazine and the solution was incubated for 20 min. Then 2 ml 0.4 N NaOH was added to the mixture and the OD of the sample was measured at 500-530 nm. The ratio AST/ALT (de Rytis coefficient) was calculated also.

Cholinesterase (ChE) Activity

Cholinesterase (ChE) was measured on the basis on hydrolysis of acetylcholinchloride to acetic acid and choline. The alterations of pH changes the colour of the mixture which was determined spectrophotometrically used the standard kit (Felicit - Diagnosis, Ukraine).

Total Soluble Protein Concentration

Total soluble protein and hemoglobin (Hb) concentration was quantified spectrophotometrically used the kits of the Filicit-Diagnosis (Ukraine). The enzyme activities were calculated per mg protein in serum and per mg hemoglobin in RBC.

2.3. Statistical Analysis

Biochemical measurements were detected in duplicate for each sample. Simple, descriptive statistics were performed using an ANOVA to determine means (+/-SEM) (Halafyan, 2008). *P* value of 0.05 was used for determination of statistical significance in all cases. The graphs were made using the computer program EXELL. Statistical correlations between studied biochemical parameters in tested animals and tissues were calculated by the least-squares method using the computer program CURVEFIT (Version 2.10-L).

3. Results

3.1. Biological, Ecological and Economic Characteristics of Examined Black Sea Fish Species

All tested fish species were joined in four ecological groups (Svetovidov, 1964; Boltachev and Karpova, 2012).

1. *Benthic group.* Three fish species belonging to benthic group were studied: scorpion fish *Scorpaena porcus*, round goby *Neogobius melanostomus* and toad goby *Mesogobius batrachocephalus*.

Round goby and toad goby are slow swimming, carnivorous species, which abundance is at the coastal waters at the depth of 20 m. They consume invertebrates such as mollusks, worms, crustacean, and fish larvae. The fecundity of the round goby is estimated as 200-400 eggs. Scorpion fish is slow swimming predator. Its abundance is located at the depth between 0,5 -1 m to 30-40 m. Its prey includes crustacean and fish. Tested benthic fish species could be proposed as biomonitors, because they are territorial forms that live and feed on the marine bottom. For this reason they intimate contact with their environment and they are very susceptible to physical, chemical and biological changes which may be modified their blood biomarkers. They are suitable for biomonitoring studies.

2. Suprabenthic group. Red mullet belongs to suprabenthic species, it is sluggish, carnivorous form. Its food includes crustacean, worms, mollusks. The abundance is at 20-30 cm over bottom., the fecundity is estimated as 10.3-28 000 eggs. It is of interest due to their commercial significance, economic importance and extensive consumption as a food source in Black Sea.

3. Suprabenthic/pelagic group. In this group we have selected high body pickerel, it is less mobile, omnivorous species which abundance is at the coastal waters depth of 5-30 m. The fecundity of the fish is estimated as 6-73 000 eggs. It possesses several unusual characteristics because this species changes its sex during ontogenesis.

4. Pelagic group. Horse mackerel is fast swimming, predator. It migrates in long distances. The fecundity of the fish is estimated as 150-200000 eggs. Its abundance is in the depth of 40 m It is of interest due to their commercial significance, economic importance and extensive consumption as a food source in Black Sea (Figure 2).

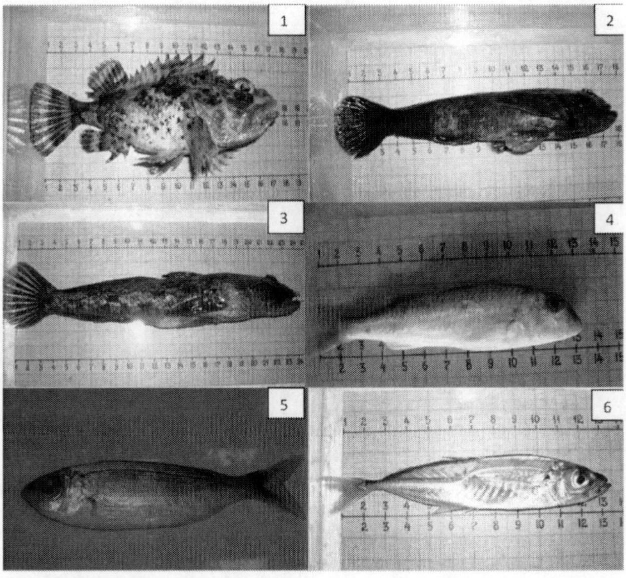

Figure 2. Scorpion fish Scorpaena porcus (1), round goby Neogobius melanostomus (2), toad goby Mesogobius batrachocephalus (3), red mullet Mullus barbatus ponticus (4), high body pickarel Spicara flexuosa (5), horse mackerel Trachurus mediterraneus (6).

3.2. Trace Elements Concentration in Fish Tissues

Concentrations of examined trace elements in fish are present in Table 1. Trace elements contents in fillets of tested Black Sea fish species were below the maximum levels permitted in fish fillets (10 for Cu, 1 for Pb, 0,2 for Cd, 40 for Zn, 5 for As and 0,4 for Hg mg• kg^{-1} wet weight). Zn was the highest in tissues of all species examined, followed by As, Cu, Pb, Hg and Cd.

The concentration of several elements in fish tissues demonstrated interspecies differences. The level of Cu was significantly higher or tended to be greater in examined benthic forms as compared with suprabenthic, suprabentic/pelagic and pelagic species. Among checked animals concentrations of copper were shown the greatest in toad goby (0.66 mg kg^{-1} wet weight) while red mullet it was the lowest (0.48 mg kg^{-1} wet weight).

Among tested fish high level of Pb was observed in scorpion fish and toad goby (0.16 and 0.13 mg kg^{-1} wet weight respectively) and the lowest level was noted in red mullet (0.067 mg kg^{-1} wet weight). Concentration of Cd was great in high body pickarel (0.05 mg kg^{-1} wet weight) as compared with other examined forms in which level varied from 0.010 to 0.036 mg kg^{-1} wet weight. Zn concentration in fish tissues ranged between 3.24 and 6.87 mg kg^{-1} wet weight, the highest value was observed in toad goby (6.87 mg kg^{-1} wet weight) and the lowest content was indicated in round goby (3.24 mg kg^{-1} wet weight).

Table 1. Concentration of trace elements in fish tissues (mg kg-1 wet weight, M + SEM)

Fish species	Cu	Pb	Cd	Zn	As	Hg
Scorpion fish	0,6 ± 0,1	0,16 ± 0,04	0,036 ± 0,01	5,14 ± 1,25	0,8 ± 0,1	0,06 ± 0,005
Toad goby	0,66 ± 0,15	0,13 ± 0,03	0,018 ± 0,006	6,87 ± 2,6	3,43 ± 0,69	0,065 ± 0,004
Round goby	0,64 ± 0,073	0,088 ± 0,03	<0,01	3,24 ± 0,23	0,92 ± 0,2	0,049 ± 0,004
Red mullet	0,48 ± 0,1	0,067 ± 0,015	0,0173 ± 0,003	4,06 ± 0,23	0,66 ± 0,2	0,05 ± 0,01
High body pickarel	0,58 ± 0,08	0,084 ± 0,017	0,05 ± 0,02	4,59 ± 0,72	1,1 ± 0,29	0,068 ± 0,01
Horse mackerel	0,53 ± 0,07	0,084 ± 0,026	0,012	5,41 ± 1,21	0,71 ± 0,24	0,051 ± 0,005

In toad goby the arsenic concentration was higher (3.43 mg kg^{-1} wet weight) as compared with the other examined species (0.66 - 1.1 mg kg^{-1} wet weight). Among tested fishes the greatest mercury level was observed in high body pickarel, toad goby and scorpion fish (0.068, 0.065 and 0.06 mg kg^{-1} wet weight correspondingly) while in other examined species it varied insignificantly from 0.049 to 0.51 mg kg^{-1} wet weight.

Hence the concentration of cooper and lead was higher in benthic forms as compared with totality of examined fish species while for other tested trace elements the differences were not significant and they varied independently.

No correlations were observed between the level of Cu and Pb and other tested elements were shown. Correlations between Cd and Hg was $r = 0.49$ and between Zn and As it was $r = 0.45$ were observed. Relationships between the other tested elements were not shown.

3.3. Oligopeptides and Oxidized Protein (OP) Concentration

Oligopeptides concentration in blood serum of the examined fish species varied between 1.93 to 4.44 optical units ml^{-1} 10^{-3} mg protein. The level in toad goby and scorpion fish was significant lower ($p < 0.05$) than in other tested fish species (Figure 3). No significant differences were shown in the serum oligopeptides level in round goby, red mullet, high body pickarel and horse mackerel.

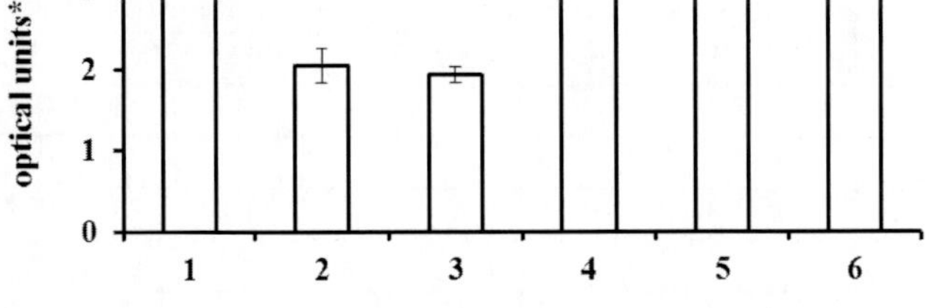

Figure 3. Oligopeptides concentration (mean ± SEM, x10^{-3} mg $protein^{-1}$) in the serum of examined fish species caught in Sevastopol Bay (Black Sea). 1- round goby, 2- toad goby, 3- scorpion fish, 4 – red mullet, 5 – high body pickerel, 6 – horse mackerel.

Serum oxidized proteins (OP) concentration ranged in tested fish species from 6.53 to 22.07 arbitrary units (Table 2). In both gobies it was approximately 2-fold higher than in other examined species. The lowest level was obtained in red mullet and horse mackerel, which were the similar, while in scorpion fish and high body pickerel the values were intermediate.

Table 2. Oxidized proteins concentration in the blood serum of examined fish species caught in Sevastopol Bay (Black Sea)

Species	Neutral products		Basic products		Σ
	ODλ346	ODλ370	ODλ 430	ODλ 530	
Round goby	5.48 ± 0.34	7.68 ± 0.49	4.62 ± 0.30	0.77 ± 0.05	18.55
Toad goby	7.14 ± 0.12	9.11 ± 0.14	5.27 ± 0.12	0.55 ± 0.04	22.07
Scorpion fish	2.70 ± 0.20	3.3 ± 0.23	1.76 ± 0.15	0.37 ± 0.04	8.13
Red mullet	2.37 ± 0.14	2.72 ± 0.15	1.52 ± 0.08	0.22 ± 0.01	6.83
High body pickerel	3.16 ± 0.25	3.57 ± 0.30	2.22 ± 0.20	0.48 ± 0.04	9.43
Horse mackerel	2.28 ± 0.30	2.40 ± 0.28	1.48 ± 0.20	0.37 ± 0.03	6.53

The ratio between the portions of different components of OP is present in Table 3. The relative content of the components registered at $OD\lambda_{370}$ was higher in the blood serum of benthic fish as compared with suprabethic, suprabenthic/pelagic and pelagic, while the values of other components varied unclear in examined species.

There is low negative correlation between the concentration of oligopeptides and oxidized proteins in tested fish blood (r = -0.39).

Table 3. The ratios of different components of oxidized proteins in fish serum (the value of $OD\ \lambda_{346}$: $OD\ \lambda_{370}$: $OD\lambda_{430}$: $OD\lambda_{530}$) where $OD\ \lambda_{346} = 1$)

Species	Neutral products		Basic products	
	ODλ346	ODλ370	ODλ 430	ODλ 530
Round goby	1	1.4	0.84	0.14
Toad goby	1	1.3	0.73	0.08
Scorpion fish	1	1.2	0.65	0.13
Red mullet	1	1.1	0.22	0.09
High body pickerel	1	1.1	0.70	0.15
Horse mackerel	1	1.05	0.21	0.16

3.4. Antioxidant Enzyme Activities

Antioxidant enzyme activities in RBC of tested fish species are present in Figure 4. SOD activity in fish RBC varied between 146.92 and 201.76 arbitrary units mg protein^{-1} min^{-1}. The highest SOD activity was shown in RBC of red mullet and the lowest value was observed in the RBC of toad goby. SOD level in horse mackerel was significantly higher ($p < 0.05$) as compared with the values of scorpion fish and toad goby (Table 4). CAT activity varied between 0.31 and 0.62 H_2O_2 mg protein^{-1} min^{-1}.The lowest CAT activity was obtained in horse mackerel, while the highest was shown in toad goby.

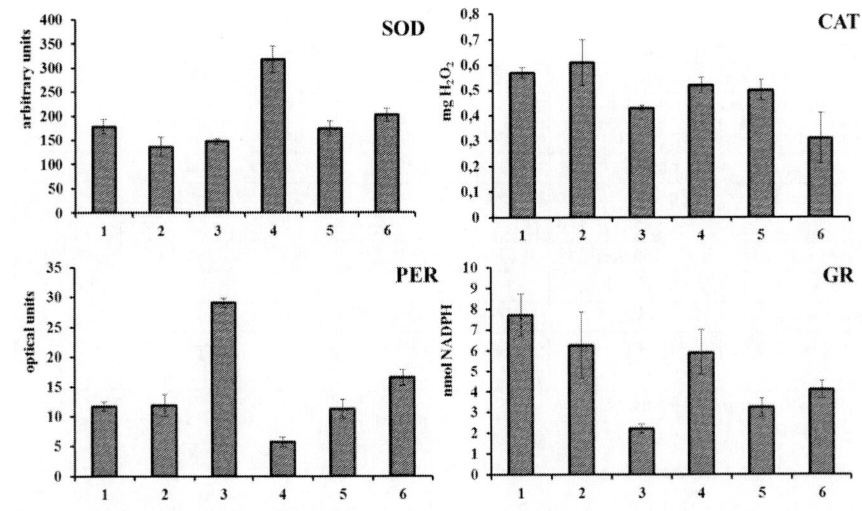

Figure 4. Antioxidant enzyme activities (mean ± SEM, x mg protein^{-1} min^{-1}) in the RBC of examined fish species caught in Sevastopol Bay (Black Sea). Other explanations see in Figure 3.

In other tested species the values varied insignificantly (Table 4). The greatest PER activity was obtained in scorpion fish (29.04 optical units mg protein^{-1} min^{-1}) and the lowest value was shown in red mullet (5.73 optical units mg protein^{-1} min^{-1}). In both gobies and high body pickerel enzyme activity was the similar, while the activity in the RBC of horse mackerel was significant higher ($p < 0.001$). GR activity was the lowest in scorpion fish (2.22 nmol mg protein^{-1} min^{-1}) and the greatest value was obtained in round goby (7.8 nmol mg protein^{-1} min^{-1}). No significant differences were shown in

RBC of toad goby and red mullet, while the enzyme activity tended increase in RBC of high body pickerel and horse mackerel as compared with scorpion fish and decrease related to the values of both gobies and red mullet.

Table 4. The differences between the values of antioxidant enzyme activities in the RBC of examined fish species

Fish species	SOD	CAT	PER	GR	GST
Round goby → toad goby	n/s	n/s	n/s	n/s	n/s
Round goby → scorpion fish	p <0.05	p<0.001	p<0.001	p<0.001	p<0.001
Round goby → red mullet	p<0.001	n/s	p<0.001	n/s	p<0.05
Round goby →high body pickarel	n/s	n/s	n/s	p<0.001	p<0.001
Round goby → horse mackerel	n/s	p<0.05	p<0.01	p<0.001	p<0.01
Toad goby → scorpion fish	n/s	p<0.05	p<0.001	p<0.05	p<0.05
Toad goby → red mullet	p<0.001	n/s	p<0.01	n/s	n/s
Toad goby → high body pickarel	n/s	n/s	n/s	n/s	n/s
Toad goby → horsemackerel	p<0.05	p<0.05	p<0.05	n/s	n/s
Scorpion fish → red mullet	p<0.001	p<0.01	p<0.001	p<0.01	p<0.01
Scorpion fish→high body pickarel	n/s	n/s	p<0.001	p<0.05	n/s
Scorpion fish → horse mackerel	p<0.001	p<0.001	p<0.001	p<0.001	p<0.001
Red mullet → high body pickarel	p<0.001	n/s	p<0.01	p<0.05	n/s
Red mullet → horse mackerel	p<0.001	p<0.001	p<0.001	n/s	n/s
High body pickerel → horse mackerel	n/s	p<0.001	p<0.05	n/s	n/s

n/s – no significant differences.

In fish red blood cells GST activity varied from 55.6 ± 0.73 $nmol^{-1}min^{-1}$mg protein in toad goby to 10.99±0.73 $nmol^{l-}min^{-1}$ mg^{-1}Hb in scorpion fish (Figure 5).

The enzyme activity in scorpion fish blood was significant lower ($p < 0.01$) than in other examined fish species. GST activity in blood cells of round goby and toad goby was the similar and it was significant greater than in other species (Table 4). Thus, GST activity demonstrated the interspecies differences.

We did not find any correlations between SOD activity and other antioxidant enzymes and GST. Low correlation were shown between CAT and GR ($r = 0.28$) and CAT and GST ($r = 0.38$) and PER and GR activity ($r = 0.29$). High correlation was observed between GR and GST activity ($r = 0.61$).

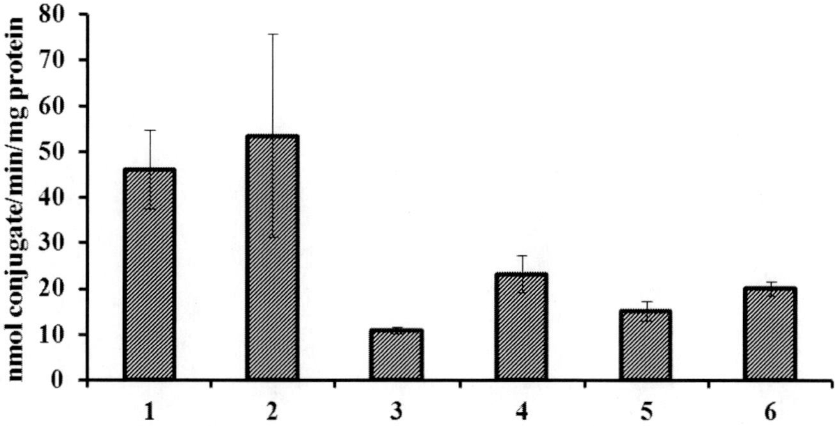

Figure 5. GST activity (mean ± SEM,) in the RBC of examined fish species caught in Sevastopol Bay (Black Sea). Other explanations see in Figure 3.

3.5. Aminotransferase Activity

Serum aminotransferase activity was significantly higher ($p < 0.05$) in toad goby as compared with other tested fish species (Figure 6). Enzyme activities were the similar in scorpion fish and red mullet. No significant differences were observed between ALT activities in RBC of round goby, scorpion fish, red mullet, high body pickerel and horse mackerel, while AST activity were higher in high body pickerel and horse mackerel as compared with round goby, scorpion fish and red mullet. De Rytis coefficient varied lower, it tended to increase in toad goby and high body pickerel (4.93 and 3.63 correspondingly), while in other checked species it was approximately the similar and varied from 2.24 to 2.62 (Figure 6).

No significant differences were shown in serum ALT activity in tested species, while in AST activity the significant differences were obtained between the values of round goby, toad goby, scorpion fish, high body pickerel and horse mackerel ($p < 0.05$); between toad goby, scorpion fish and red mullet ($p < 0.05$), between scorpion fish and horse mackerel ($p < 0.001$), between red mullet and horse mackerel ($p < 0.01$).

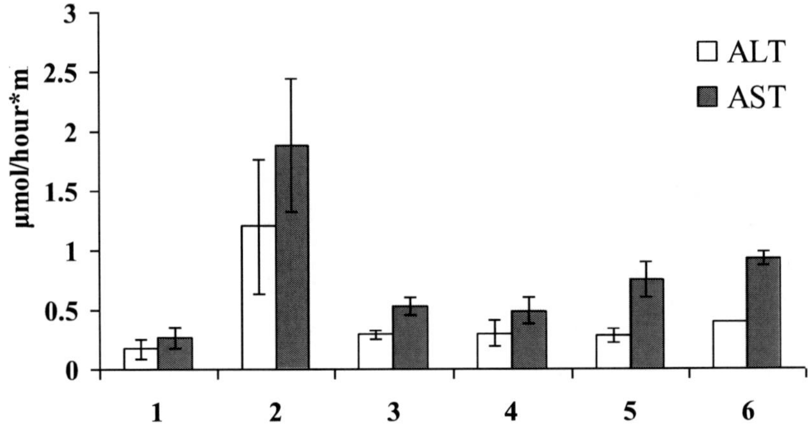

Figure 6. ALT and AST activities (mean ± SEM) in the serum of examined fish species caught in Sevastopol Bay (Black Sea). Other explanations see in Figure 3.

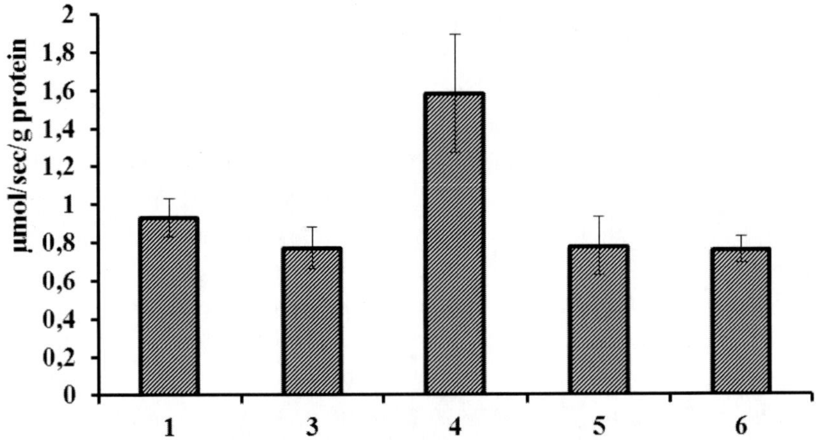

Figure 7. ChE activity (mean ± SEM,) in the RBC of examined fish species caught in Sevastopol Bay (Black Sea). Other explanation see Figure 3.

3.6. Cholinesterase Activity

The highest cholinesterase activity was determined in red mullet serum 1.58 $\pm$ 0.31 μmol sec^{-1} g $protein^{-1}$ (Figure 7). It was significant greater

($p < 0.05$) than the enzyme level in the other tested fish species. No significant differences were shown between the values of the other examined species.

Therefore, we found interspecies differences in trace elements concentration in fillet, oligopeptides and oxidized proteins content and enzyme activities in the blood of tested fish species in Black Sea.

4. DISCUSSION

Fish physiological status depends on their taxonomic position and the adaptations to habitat conditions (Martinez-Alvarez et al., 2005). The biodiversity of teleost species is very high and they comprise different ecological groups: benthic slow swimming forms, sluggish supra/benthic species, less active suprabenthic, suprabenthic/pelagic animals, and active pelagic fishes, which are very different each from other. Among tested forms there are predators (scorpion fish and horse mackerel) and they are at the top of the food chains in ecosystem, other fish species are carnivorous (round goby, toad goby and red mullet), the high body pickerel is omnivorous. Thus, general prey categories, trophic levels and feeding behavior are also differed between the examined species, selected in our study.

There are several factors including abiotic (physical and chemical parameters of the water and sediments in marine location, water temperature, pH, salinity, and etc.), biotic (specificity of fish biology, feeding behavior, age, sex, swimming activity and metabolic rate) and anthropogenic (level of pollution caused human activity) that may change fish physiology and biochemistry (Sole et al., 2009, 2010; Barrera-Garcia et al., 2012; Mathews and Fisher, 2009; Velez-Alavez et al., 2013). Different level of pollution in biotopes caused the different accumulation of pollutants, including trace elements in fish tissues. Previously we showed the fluctuation of trace elements and nitrosamine levels in Black Sea elasmobranchs which could be associated with the interspecies variations in diet and living conditions, when rays are principally benthic forms whereas the shark is bento-pelagic predator (Rudneva et al., 2012b; 2014).

Taking into account the specificity of the biology and ecology of tested fish species the aim of the present study was to compare their biochemical characteristics and to indicate their differences in blood biomarkers, because blood biochemistry is important in evaluation of fish physiological and ecological status and health.

Hence, the discussion of the results obtained in present study will consider four main points:

- to compare the trace metals concentrations in muscle and examined biochemical indicators in the blood of six selected Black Sea fish species, belonging to different ecological groups;
- to show the interspecies differences in tested trace elements concentrations and blood biochemical indicators;
- to indicate the correlation between blood biomarkers and trace elements concentration in fish tissues;
- to explain possible causes of the interspecies differences of examined indicators related to fish ecological status.

4.1. Comparison of Trace Elements Concentrations in Fish Muscle

Six trace elements concentrations in the fillet of examined Black Sea fish species were comparable to those reported elsewhere for fish. In the muscle tissues of tested animals the concentrations of checked elements was Zn > As > Cu > Pb > Hg > Cd. The concentration of trace metals in samples is depended on fish species.

Comparing the examined species, Cu concentration in the tissues of checked Black Sea fishes caught in Sevastopol Bay, ranged between 0.48 and 0.66 mg kg^{-1} wet weight. The level was significantly higher in benthic forms (scorpion fish, toad goby and round goby) as compared with the other species. The next essential element Zn level ranged in the tissues of examined Black Sea species between 3.24 and 6.87 mg• kg^{-1} wet weight. The highest concentration was shown in toad goby, while the lowest content was indicated in round goby. Therefore, concentration of these two essential elements was comparable with the content in the muscle of the fish species from different geographical locations which ranged between 0.2 and 18.3 mg• kg^{-1} wet weight (Turoczy et al., 2000) including Black Sea also (Mendil et al., 2010).

Pb level was significantly higher ($p < 0.01$) in two gobies and scorpion fish than in other examined animals. The highest Cd level was indicated in high body pickarel. It was more than 1.5-2-fold greater than in other investigated species. Mercury level varied lower in Black Sea fish species (0.49-0.68 mg• kg^{-1} wet weight). The concentration of the mercury in the muscle of examined species caught in Sevastopol Bay was comparable with the values founded in the fillets of the fish from Izmir Bay (Eastern Aegan, Turkey), ranging from 0.023 to 0.224 mg• kg^{-1} wet weight (Gonul and Kucuksezgin, 2005).

Fish have been recognized as the most important source of mercury in aquatic environments while invertebrates (mollusks and crustacean) are the important source for cadmium (Marcovecchio et al., 1991). Trace metal accumulation in teleost species tested from Sevastopol Bay may thus be expected since they are in their corresponding trophic net. Dietary differences among species seemed to be the most important causes for differences in their trace elements level in muscle.

Arsenic was significantly greater in toad goby compared to other checked species. Feeding habits have important influence on arsenic levels in marine fish: animal feeding of algae and crustacean appear to retain higher arsenic concentrations than piscivorous forms. In this case the differences of As concentrations in tested species might be related to diet because some of them are predators (scorpion fish and horse mackerel), the other are carnivorous (both gobies) and omnivorous (red mullet and high body pickerel).

The levels of tested trace elements content obtained in Black Sea fish species caught in Sevastopol Bay were comparable with the data obtained investigators for Black Sea fish species. The maximum metal concentrations were found to be as 17.8–25.7 μg/g (Zn), 0.28–0.64 μg/g (Pb), 1.4–1.9 μg/g (Cu), 0.18–0.35 μg/g (Cd) and) for fish species. *Sarda sarda*, *Mullus barbatus ponticus*, *Trachurus trachurus* and *Merlangius merlangus* collected from the Black Sea, Turkey between 2008 and 2009. The authors conclude that in spite of some species is accumulated trace metals at high ratio, the levels in analyzed fish were acceptable to human consumption at nutritional and toxic levels. However, the levels of lead and cadmium in fish samples were higher than the recommended legal limits (Mendil et al., 2010). In another publication concentrations of As, Cd, and Pb in scale, skin, muscle, gills, liver and the gonads of gilthead seabream (*Sparus aurata*), European seabass (*Dicentrarchus labrax*), Black Sea salmon (*Salmo trutta labrax*), horse mackerel (*Trachurus trachurus*), red mullet (*Mullus barbatus*) and whiting (*Merlangius merlangius euxinus*) from the Black Sea, Turkey the mean metal concentrations in different tissues were in the following ranges: As 0.13-3.40, Cd 0.32-6.25, Pb 0.02-0.38 mg kg^{-1} wet weight. Metal concentrations in the muscles of the examined species were generally lower than those in scale, skin, gills, liver and the gonads (Aydın D. and Tokalıoğlu Ş., 2015). Hence the trace element levels obtained in the tissues of Black Sea fish species caught in

Sevastopol shelf were compared to those which have been reported in fish worldwide and in the Black Sea.

There are several factors including abiotic (physical and chemical parameters of the water and sediments in marine location), biotic (specificity of fish biology, feeding behavior, age, sex, swimming activity and physiological processes) and anthropogenic (level of pollution led human activity) that may affect trace elements accumulation in marine organisms. The most important factor is diet. It is well known that animals that include a large quantity of fish in their diet have the tendency to accumulate mercury to higher concentrations while higher cadmium level was found in fish in which diet invertebrates predominate (Storelli et al., 2005).

Several researchers have reported trace element concentrations associated with body size in marine fish (Turoczy et al., 2000; Gonul and Kucuksezgin, 2005). Strong positive correlation between fish size and metal content (mercury, copper and cadmium) were obtained for fish (Marcovecchio et al., 1991). Data from the literature emphasis the importance of biometric parameters in studies on the bioaccumulation of mercury (Storelli et al., 2005). Similar results were documented for Hg concentrations in muscle and the length of teleost species from Izmir Bay (Gonul and Kucuksezgin, 2005). The results reflect the increase of trace elements accumulation with age and body size. As a rule in, older fish are larger animals and they require large prey for feeding. Large feeding organisms contain high levels of xenobiotics, including trace elements which are accumulated in the fish body (Turoczy et al., 2000). In contrast, several authors reported that metal contents tended to decrease with fish length which was documented at the case of deep-water sharks caught in North-eastern Atlantic (Vas et al., 1993).

We could suggest that both body size and the belonging of the fish to the specific ecological group cause the trace elements level in the muscle. Among tested elements mercury, lead, cadmium and arsenic are non-essential elements while Cu and Zn are essential. Among the fish examined there are the forms with the different feeding behavior and they can accumulate different concentrations of the toxicants. Additionally fishes demonstrate different metabolic rate because pelagic and suprabenthic/pelagic forms are typically active swimming species and benthic fish are sluggish forms. High metabolic rates increases food requirements which contributes notably to the accumulation of pollutants (Storelli et al., 2005).

Obtained results demonstrated the interspecies differences between checked fish. Content of Cu, Pb, As and Hg was higher in majority of benthic forms caught in Sevastopol Bay. It could be associated with the interspecies differences in diet and living conditions. Scorpion fish and both gobies are principally benthic forms whereas the others are bento-pelagic and pelagic animals. Higher concentrations of the trace metals in benthic species may result from increased exposure to them due to the ingestion of toxicant-containing sediments and food (Goksoyr et al., 1996; Porte et al., 2000; Da Rocha et al., 2009; Sole et al., 2009). High level of pollution in Sevastopol coastal waters influences negatively on tested fish species and stimulates oxidative stress, damage of biomolecules and their accumulation in tissues and organs.

4.2. Comparison the Biomarkers of Damage (Oligopeptids and Oxidized Proteins Concentration) in Fish Serum

Trace elements are concentrated in to the fish in the marine food nets and causes oxidative stress in fish directly and indirectly (Livingstone, 2001; Barata et al., 2005). Reactive oxygen species (ROS) production increases and damages biomolecules, membranes and cells. Lipids and proteins are very sensitive to ROS and they are the main targets of the ROS attack. At the other way, the biotransformation of xenobiotics and ROS generation in organism are associated with the formation of low molecular weight components, such as oligopeptides.

Oligopeptides play a role in organism metabolism, involved in intercellular gene signaling processes. They also modulate biochemical pathways and some of them mimic regulatory peptides, they can bind with cell receptors and change cell response. Level of oligopeptides depends on proteasomal system activity and the balance between the formation and excretion of toxins and their intermediates (Grune, 2000). Increase of oligopeptides level in the tissues is associated with the damage of the balance between protein synthesis and degradation, decrease of proteasomal function or both these processes (Vysotskaya and Nemova, 2008).

The lowest concentration of oligopeptides was detected in blood serum of toad goby and scorpion fish, while the values in other examined species were significantly higher (approximately 2-fold). Oligopeptides may form during protein degradation in the liver as well as during biotransformation of xenobiotics, generating different kinds of low molecular weight molecules. High level of oligopeptides in fish serum could be explained by transport and excretory function of blood serum, because many intermediates and metabolites, including different kinds of oligopeptids, are present in it.

Great level of oligopeptides in the serum of suprabenthic, suprabenthic/pelagic and pelagic forms could be associated with their active swimming and feeding behavior, while round goby and scorpion fish are more sluggish and their metabolic rate is lower. However, the concentration of oligopeptides in toad goby was significantly higher as compared with the other benthic species. We could proposed that it is specificity of the metabolism of this species. The data of these compounds concentration in fish tissues are very limited and further investigations are needed for understanding the interspecies differences and the influence both biotic and abiotic factors on their values. We could suggest, that the obtained changes could be associated with the specificity of metabolic rate of the checked species and the conditions of biotops, probably with the depth. Taking into account, that among oligopeptides there are many biologically active compounds, we could propose also that they are used as intermediates in different metabolic pathways in tested fish species as we described previously in Black Sea elasmobranches (Rudneva et al., 2014).

Our findings demonstrated the increasing of oxidized proteins in blood serum of both gobies, while in scorpion fish it was lower and comparable with the animals belonging to the other ecological groups. As we described previously, there are several factors including abiotic, biotic and anthropogenic that may change fish physiology and biochemistry (Sole et al., 2009; Barrera-Garcia et al., 2012; Mathews and Fisher, 2009; Velez-Alavez et al., 2013). Oxidative stress caused unfavorable living conditions is an imbalance toward the pro-oxidant side of the pro-oxidant/antioxidant homeostasis. The content of protein carbonyl groups are good biomarkers of oxidative stress because they have some advantages in comparison with the measurement of other oxidation products: they are the relative early formation and the relative stability of carbonylated (oxidated) proteins (Dalle-Donne et al., 2003; Luschak, 2007). Carbonyl (CO) groups (aldehydes and ketones) are produced on protein side chains when they are oxidized. These moieties are chemically stable, therefore protein carbonyl content is the most general indicator of protein oxidation. The accumulation and increase of carbonyls in proteins has been observed in several human diseases (Dalle-Donne et al., 2003; Luschak, 2007). Oxidative damage on proteins may lead to negative consequences because among the proteins there are enzymes, hormones, mediators and other biomolecules which play an important role in metabolism and homeostasis of the organism.

Our results show that oxidized proteins concentration increased in the blood serum of two bottom fish species caught in marine polluted area. We could propose that pollutants stimulate oxidative stress in fish. High level of ROS causes inducible prooxidants overproduction and protein carbonylation. However, our findings demonstrated the interspecies differences in serum oxidized proteins level: in scorpion fish the concentration was lower than in both gobies. Thus we could propose that scorpion fish is more resistant to pollution as compared with both gobies and the concentration of oxidized proteins depends of its feed behavior. The scorpion fish is predator, while gobies are carnivorous. Thus we could propose the differences in the protein metabolism of the fish and the accumulation of oxidized proteins in the tissues and in blood serum. Previously we also demonstrated the interspecies differences of oxidized proteins concentration between several Black Sea teleosts and their relations to heavy metals concentration in fish tissues (Rudneva et al., 2011). We indicated that high oxidized proteins level correlated with the increase of. concentrations of trace elements in fish tissues, several of which are well known as prooxidants, caused overproduction of ROS. We also demonstrated the seasonal dynamics of serum oxidized proteins concentration and its relation to antioxidant enzymes activity in the blood of round goby in Black Sea and the Sea of Azov, and we showed that the tested parameters depended on fish physiological status (Kovyrshina and Rudneva, 2012). Our findings demonstrated also that benthic species are not uniform and complex of factors influenced on their physiological status.

We found negative low correlation between the oligopeptides and oxidized proteins concentration in fish serum ($r = -0.39$), which demonstrated the connection between the generation of oligopeptides and protein damage in fish tissues. High correlations were observed between the concentration of heavy metals Pb and Hg and oligopeptides content in fish ($r = 0.82$ and $r = 0.52$ correspondingly). Direct correlations were shown between Cu and As levels in fish tissues and the concentration of oxidized proteins in blood serum ($r = 0.65$ and $r = 0.51$ respectively). Therefore, trace elements accumulation may change fish protein metabolism which depends on totality ecological and anthropogenic factors in Black Sea. At the other hand, species belonging to the similar ecological groups are not uniform and they are differed each from other.

4.3. Comparison the Biomarkers of Resistance against Oxidative Stress

Organisms have antioxidant defense to counteract the toxic effects of ROS. It includes special adaptive enzymes superoxide dismutase (SOD), catalase (CAT), peroxidase (PER), glutathione reductase (GR). SOD protects against oxidative damage by catalyzing the reaction of dismutation of the superoxide anion to H_2O_2, that degrades by CAT to H_2O. PER includes peroxide-degrading enzymes both specific and nonspecific, that reduce both hydrogen peroxide and hydroperoxides. GR maintains a ratio of GSH/GSSG under oxidative stress, it is a major regulatory enzyme that regenerates GSH from GSSG (Livingstone, 2001).

Our findings demonstrated the interspecies differences in RBC antioxidant activities in examined fish, but we could not note the relationships between these parameters and specificity of fish ecology. The interspecies variations of antioxidant enzyme levels were shown between fish species and it was clearly quite different. (Hamed et al., 2003; 2004; Martinez-Gomez et al., 2005). It depends on the peculiarities of their phylogenetic position, biology, physiology and ecological status of the locations (Martinez-Alvarez et al., 2005). GST activity in blood cells of all examined fish species varied insignificantly with the exception of both gobies in which that was significant greater. Thus the study of antioxidant enzyme activities in six Black Sea fish species allowed us to identify differences between them. We could not suggest that the ecological status influences on the antioxidant enzyme activity in blood because pelagic horse mackerel, suprabenthic and suprabenthic/pelagic high body pickarel and red mullet, benthic round goby, toad goby and scorpion fish showed differences of enzyme activity. We could not demonstrate the correlation between fish ecological status and enzyme activities in blood cells also.

At the other side the interspecies variations of examined Black Sea fish species may be the result of the different fish sensitivity to metal and organic pollution. Many investigators demonstrated the increasing of antioxidant enzyme activities in hepatopancreas in aquatic invertebrates and in the blood in fish, inhabited high polluted sites. The tissue specific damage corresponded to the differences in the antioxidant enzyme activity potentials of the tissues for their adaptation to environmental stress (Ahmad et al., 2004; Rudneva et al., 2010 a, b). Previously we described the high anthropogenic impact in Sevastopol Bay (Black Sea) and its negative consequences in fish health (Rudneva and Petzold-Bradley, 2001). The interspecies variations of antioxidant enzyme activity in fish tissues may reflect the specific adaptations to the oxidative stress and protective mechanisms against oxidative damage. However, we did not find high correlations between the RBC antioxidant enzyme activities and the level of trace elements in fish tissues with the exception of the relationships between SOD activity and Cu content (r = 0.69), PER activity and Pb level (r = 0.57), GR activity and Zn level (r = 0.33). Low correlation was shown between GST activity and Cu (r = 0.23) and Cd level (r = 0.17).

Thus, the analysis of these biomarkers in fish organs and tissues is important for the evaluation of fish abilities to protect against organic pollution and keep their life in the polluted environments.

4.4. Comparison of Fish Health Biomarkers

An increase of aminotransferase enzyme activity in the blood serum, plasma and other extracellular fluids was shown in the tissues of the organisms impacted unfavorable conditions, organ dysfunction or internal lesions in tissues. The damaged cells release their contents (including aminotransferases) towards the blood stream, and the level of these enzymes enhances in serum (Martinez-Porchas et al., 2011). AST is used as clinic diagnostic tool, and it is associated with cell necrosis of the liver and skeletal or cardiac muscle, starvation and lacking of vitamin E. Plasma ALT is an acute hepatic damage good marker (Coppo et al., 2001-2002).

In our study the interspecies variations of serum aminotransferase enzyme activities demonstrated the high levels of AST as compared with ALT in the serum of examined fish species. At the other hand we could not connect these differences with fish ecological group directly. Enzyme activities in benthic forms was not uniform because the highest value was shown in toad goby while the activity of the other checked species was comparable with the suprabenthic and pelagic forms. However, AST activity was higher in suprabenthic/pelagic and pelagic species as compared with benthic and suprabenthic forms with the exception on toad goby. We propose that these interspecies differences may depend on fish ecological status; benthic and suprabenthic forms are generally slow swimming forms, and suprabenthic/pelagic and pelagic species are more active animals. An increase of aspartat aminotransferase activity in the active fish may be explained by the high metabolic rate and swimming activity of this species, while these benthic forms have lower metabolic rates.

Pollution level also may modulate the biochemical characteristics of the tested animals belonging to benthic, suprabenthic, suprabenthic/pelagic and pelagic biotops of Black Sea. We found high correlations between ALT and AST activity in serum and level of Zn ($r = 0.69$ and $r = 0.81$ respectively) and As ($r = 0.89$ and $r = 0.77$ respectively) in fish muscle. Fish living in marine locations contaminated with different xenobiotics, may be exposed and stressors caused a variety of toxic intermediates. Analysis of aminotransferase activities in fish tissues is important for the evaluation of fish abilities to protect against harmful factors and to keep their life in the pollute environments.

Serum ChE activity was the highest in red mullet, while in other tested fish species the enzyme activity was approximately the similar. We did not found any correlation between the content of examined trace elements in fish fillets and ChE activity in the blood serum. However, decrease of enzyme activity may reflect the pollution of organophosphate chemicals, metals, detergents which were contained in domestic sewage and maritime effluents (Van der Oost et al., 2003; Napierska et al., 2009). ChE inhibition by organophosphorus (OP) compounds follows different behaviors depending on pesticide chemical structure. Ions can alter ChE activity inhibiting or activating so that it was proposed that enzymes could be used as biomarkers of heavy metals pollution. Another concern about application fish ChEs as biomarker of pesticides is that in eutrophed sites some species of cyanobacteria produce anticholinesterasic metabolites such as anatoxin-a which can be considered natural OP compounds and which toxicity can be approximately 1000-fold higher that of the insecticide paraoxon (Assis et al., 2011). Cholinesterases may reflect the disturbance of neural functions of the organism in stress conditions. Monitoring of ChE inhibition has been measured in the tissues of marine fish seems to be a valid indicator for chemical pollution of the marine environment as well (Kirbi et al., 2000). Our previous study demonstrated the differences of ChE activity in scorpion fish caught in different polluted areas (Rudneva et al., 2012a).

Thus, the examined biochemical indicators in fish RBC and serum may reflect fish health and ecological status of the environment and they could lead to an early detection of ecotoxicity in the waters in different depth.

Conclusion

Blood biomarkers of the tested Black Sea fish species may provide information about interspecies differences and the mechanisms of adaptation of each species to their environment conditions. Moreover, six tested teleosts may be considered suitable for understanding the mechanisms of the adaptation of marine organisms to variety of ecological conditions in marine ecosystems and to metabolic requirements depending on their natural activity and behavior. A number of environment factors both biotic and abiotic influence on fish metabolic rate and their biochemical characteristics in the species belonging to different ecological groups.

A lot of people focus their attention on environmental problems of Black Sea and especially on its coastal areas which are the main recipients of sewage. Basically, restoration of Black Sea ecosystem is possible using the optimal ecosystem management and collaboration between riparian and non-riparian states for protecting of Black Sea environment. At this case monitoring studies and application the ecotoxicological approach may be helpful for the evaluation of ecosystem status. Fish are good monitors and biomarkers may use successfully in monitoring programs. The values of the examined biochemical parameters of fish blood and serum obtained in this study could be used as biomarkers to evaluate the fish health and changes in water quality caused different stressors and pollutants.

References

Ahmad, I., Pacheco, M. and Santos, M. A. (2004). Enzymatic and nonenzymatic antioxidants as an adaptation to phagocyte-induced damage in *Anguilla anguilla* L. following in situ harbor water exposure. *Ecotoxicol. Environ. Saf.*, 57, 290-302.

Amiard-Triquet, C. and Berthet, B. (2015). Individual biomarkers. In C. Amiard-Triquet, J-C. Amiard and C. Mouneyrac (Eds.), *Aquatic Ecotoxicology: Advancing Tools for Dealing with Emerging Risks*. (pp. 153-182). USA: Academic Press.

Assis, C. R. D., Bezerra, R. S. and Carvalho, L. B. (2011). Fish cholinesterases as biomarkers of organophosphorus and carbamate pesticides. *In* M. Stoytcheva (Ed), *Pesticides in the Modern World – Pests Control and pesticides Exposure and Toxicity Assessment* (pp. 254-278). USA: InTech.

Aydın, D. and Tokalıoğlu, Ş. (2015). Trace metals in tissues of the six most common fish species in the Black Sea, Turkey. *Food Addit. Contam. Part B Surveill.*, 8(1), 25-31.

Barata, C., Lecumberri, I., Vila-Ascale, M., Prat, N. and Porte, C. (2005) Trace metal concentration, antioxidant enzyme activities and susceptibility to oxidative stress in the tricoptera larvae *Hydropsyche exocellata* from the Lobregat river basin (NE Spain). *Aquat. Toxicol.*, 74(1), 3-19.

Barrera-Garcia, A., O'Harra, T., Galvan-Magana, F., Mendez-Rodriguez, L. C., Castellini, M. J. M. and Zenteno-Savin, T. (2012). Oxidative stress indicators and trace elements in the blue shark (*Prionace glauca*) of the east coast of the Mexican Pacific Ocean. *Comp. Biochem. Physiol. Part C*, 156, 59-66.

Boltachev, A. R. and Karpova, E. P. (2012). *Marine fishes of the Crimean Peninsula*. Simpheropol: Business –Inform. (in Russian).

Coppo, J. A., Mussart, N. B. and Fioranelli, S. A., (2001-2002). Physiological variation of enzymatic activities in blood of bullfrog, *Rana catesbeianna* (Shaw, 1802). *Rev. Vet.*, 12/13 (1-2), 22-27.

Da Rocha, A. M., Salomao de Freitas, D. P., Burns, M., Vieira, J. P., de la Torre, F. R. and Monserrat, J. M. (2009). Seasonal and organ variations in antioxidant capacity, detoxifying competence and oxidative damage in freshwater and estuarine fishes from Southern Brazil. *Comp. Biochem. Physiol. Part C*, 150, 512-520.

Dalle-Donne, I., Rossi, R., Giostarini, D., Milzani, A. and Colombo, R. (2003). Protein carbonyl groups as biomarkers of oxidative stress. *Clin. Chim. Acta*, 329, 23-38.

Di Giulio, R. T., Washburn, P. C., Wenning, R. J., Winston, G. W. and Jewell, C. S. (1989). Biochemical responses in aquatic animals: a review of determination of oxidative stress. *Environ. Toxicol. and Chem.*, 8, 1103-1123.

Dubinina, E. E., Burmistov, S. O., Hodov, D. A. and Porotov, I. G. (1995). Oxidative modification of human serum proteins, the method of determining. *Questions of medicine chemistry*, 1, 24-26. (*in Russian*).

Farombi, E. O., Adelowo, O. A. and Ajimoko, Y.R. (2007). Biomarkers of oxidative stress and heavy metal levels as indicators of environmental pollution in African cat fish (*Clarias gariepinus*) from Nigeria Ogun River. *Environ. Res. Public Health*, 4(2),158-165.

Fazio, F., Faggio, C., Marafioti, S., Torre, A., Sanfilippo, M. and Piccione, G. (2013a). Effect of water quality on heamatological and biochemical parameters of Gobius niger caught in Faro lake (Sicily). *Iran J. fish. Sci.*, 12 (1), 219-231.

Fazio, F., Faggio, C., Marafioti, S., Torre, A., Sanfilippo, M. and Piccione, G. (2012). Comparative study of haemotological profile on Gobius niger in two different habitat sites: Fario Lake and Tyrrhenian Sea. *Cah. Biol. Mar.*, 53, 213-219.

Fazio, F., Marafioti, S., Arfuso, F., Piccone, G. and Faggio, C. (2013b). Comparative study of the biochemical and haematological parameters of four wild Tyrrhenian fish species. *Vet. Med.*, 58 (11), 576-581.

Fazio, F., Marafioti, S., Torre, A., Sanfilippo, M., Panzera, M and Faggio, C. (2013c). Haematological and serum protein profiles of Mugil cephalus: effect of two different habitats. *Ichthyol. Res.*, 60, 36-42.

Goksoyr, A., Beyer, J., Egaas, E., Grosvik, B. E., Hylland, K., Sandvik, M. and Skaare, J. U. (1996). Biomarker responses in flounder (*Platichthys flesus*) and their use in pollution monitoring. *Mar. Pollut. Bull.*, 33(1-7), 36-45.

Goldberg, D. M. and Sparner, R. J. (1987). Glutathione reductase. In: H.U. Bergmeyer, J. Bergmeyer and M. Grab (Eds). *Methods of Enzymatic Analysis. Vol. III. Verlag Chemic* (258-265). Weinheim.

Gonul, L. T. and Kucuksezgin, F. (2007). Mercury accumulation and speciation in the muscle of red mullet (*Mullus barbatus*) and annular sea bream (*Diplodus annularis*) from Izmir Bay (Eastern Aegen). *Mar. Pollut. Bull.*, 54, 1962-1967.

Grune, T. (2000). Oxidative stress, aging and the proteasomal system. *Biogerontology*, 1, 31-40.

Habig, W. H., Pabst, M. J. and Jokoby, W.B. (1974). Glutathione S-transferase. The first enzyme step in mercapturic acid formation. *J. Biol. Chem.*, 249, 7130-7139.

Halafyan, A. A. (2008). *Data statistic processing. Analysis of Variance. In STATISTICA 6* (pp. 133-152). Moscow: Binom. (*in Russian*).

Haman, K. H., Norton, T. M., Thomas, A. C., Dove, A. D. M. and Tseng F. (2012). Baseline health parameters and species comparisons among free-ranging Atlantic sharpnose (*Rhizoprionodon terraenovae*), bonnethead (*Sphyrna tiburo*) and spiny dogfish (*Squalus acanthias*) sharks in Georgia, Florida, and Washington, USA. *J. Wildlife Diseases*, 48(2), 205-306.

Hamed, R. R., Farid, N. M., Elowa, Sh. E. and Abdalla, A.-M. (2003). Glutathione related enzyme levels of freshwater fish as bioindicators of pollution. *The Environmentalist*, 23, 313-322.

Hamed, R. R., Maharem, M. and Guinidi, R. A. M. (2004). Glutathione and its related enzymes in the Nile Fish. Fish. *Physiol. Biochem.*, 30, 189-199.

Hansson, T., Schiedek, D., Lehtonen, K. K., Vuorinen, P. J., Liewenborg, B., Noaksson, E., Tjarnlund, U., Hansson, M. and Balk, L. (2006). Biochemical biomarkers in adult female perch (*Perca fluviatilis*) in a chronically polluted gradient in the Stockholm recipient (Sweden). *Mar. Pollut. Bull.*, 53, 451-468.

Hotard, S. and Zou, E. (2008). Activity of Glutathione-S-transferase in the hepatopancreas is not influenced by the molting cycle in the Fidder crab *Uca pugilator. Bull. Environ. Contam. Toxicol.*, 81, 242-244.

Karyakina, E. V. and Belova, S. V. (2004), Short sequence peptides as an integral indicator of metabolic disorders (review). *Clin. Lab. Diag.*, 3, 3-8. (*in Russian*).

Konovalov, S. M. (1996). Introduction. In S. M. Konovalov (ed.), *The modern state of Black Sea ichthyophauna* (pp. 7-8). Sevastopol: Institute of the Biology of the Southern Seas. (*in Russian*).

Kovalev, A. V. and Gubanova, A. D. (1995). Long-term dynamics of Sevastopol Bay plankton. In V. N. Eremeev (Ed.), *Investigations of the Shelf Zone of the Azov-Black Sea Basin.* (pp. 96-99). Sevastopol: Proceedings of Marine Hydrophysical Institute. (*In Russian*).

Kovyrshina, T. B. and Rudneva, I. I. (2012). Seasonal dynamics of activity of oxidative modification of proteins and oxidation-inhibiting enzymes in the blood of goby *Neogobius melanostomus* inhabiting the Black Sea and the Sea of Azov. *J. Ichthyology*, 52 (4), 277-283.

Lane, N. (2011). The Evolution of Oxidative Stress. In K. Pantopoulos and H. M. Schipper (Eds.), *Principles of Free Radical Biomedicine. Volume 2.*(pp. 1-17). Nova Science Publishers, Inc.

Litvin, F. F. (1981). Laboratory manual of physicochemical methods in biology. Moscow: Moscow State University.

Livingstone, D. R. (2001). Contaminant-stimulated reactive oxygen species production and oxidative damage in aquatic organisms. *Mar. Pollut. Bull.*, 42, 656–665.

Lowry, O. H., Rosebrough, N. J., Farr, A. L. and Randall, R. J. (1951). Protein measurement with the Folin-phenol reagent. *J.Biol.Chem.*, 193, 165-175.

Luschak, V. I. (2007). Free radical protein oxidation and its link with organism functional status. *Biochem.*, 72(8), 995-1017. (*in Russian*).

Marcovecchio, J. E., Moreno, V. J. and Perez, A. (1991). Metal accumulation in tissues of shark from the Bahia Blanca estuary, Argentina. *Mar. Environ. Res.*, 31(4), 263-274.

Martinez-Alvarez, R., Morales, A. and Sanz, A. (2005). Antioxidant defenses in fish: biotic and abiotic factors. *Rev. Fish Biology and Fisheries*, 15, 75-88.

Martinez-Gomez, C., Fernandez, B., Valdes, J., Campillo, J. A., Benendicto, J., Sanchez, F. and Vethaak, A. D. (2009). Evaluation of three-year monitoring with biomarkers the *Prestige* oil spill (Spain). *Chemosphere*, 74(5), 613 – 620.

Martinez-Porchas, M., Hernandez-Rodriguez, M., Davila-Ortiz, J., Vila-Cruz, V. and Ramos-Enriquez, J. R. (2011). A preliminary study about the effect of benzo[α]pyrene (BaP) injection on the thermal behavior and plasmatic parameters of the Nile tilapia (*Oreochromis niloticus* L.) acclimated to different temperatures. *Pan-American J. Aquatic Sciences*, 6(1), 76-85.

Mathews, T. and Fisher, N.S. (2009). Dominance of dietary intake of metal in marine elasmobranchs and teleost fish, *Sci. Total Environ.*, 407, 5156-5161.

Mendil, D., Demirci, Z., Tuzen, M. and Soylak, M. (2010). Seasonal investigation of trace element contents in commercially valuable fish species from the Black sea, Turkey. *Food Chem. Toxicol.*, 48 (3), 865–870.

Napierska, D., Barsiene, J., Mulkiwicz, E., Podolska, M. and Rybakovas, A. (2009). Biomarker responses in flounder *Platichthys flesus* from the Polish coastal area of the Baltic Sea and applications in biomonitoring. *Ecotoxicol.,* 18, 846-859.

Nishikimi, M., Rao, N.A. and Yagik, K. (1972).The occurance of superoxide anion in the reaction of reduced phenazine. *Biochem. Biophys. Res. Commun.*, 46 (2), 849 -854.

Porte, S., Escartin, E., Garcia, L.M., Sole, M. and Albaiges, J. (2000). Xenobiotic metabolizing enzymes and antioxidant defences in deep-sea fish: relationship with contaminant body burden. *Mar. Ecol. Prog. Series.*, 192, 259-266.

Rudneva, I.I., Kuzminova, N.S. and Skuratovskaya, E.N. (2010a). Glutathione-S-transferase activity in tissues of Black Sea fish species. *Asian J. Exp.Biol. Sci.*, 1(1), 141-150.

Rudneva, I. I., Dorokhova, I. I., Skuratovskaya, E. N. and Kuzminova, N. S. (2014). Comparative studies of hepatic and blood biomarkers of three species of Black Sea elasmobranchs. *Int. J. Mar. Sci.*, 4(44), 1-14.

Rudneva, I. I., Skuratovskaya, E. N., Dorohova, I. I. and Kovyrshina, T. B. (2012a). Use of fish blood biomarkers for evaluating of marine environment health. *World J. Sci. Technol.*, 2(7), 19-25.

Rudneva, I.I. (1997). Blood antioxidant system of Black Sea elasmobranch and teleosts. *Comp. Biochem. Physiol.*, 118C, 255-260.

Rudneva, I.I. (2011). *Ecotoxicological Studies of the Black Sea Ecosystem. The Case of Sevastopol Region*. New York: Nova Science Publishers.

Rudneva, I.I. (2012). Antioxidant defense in marine fish and its relationship to their ecological status. In S. P. Dempsy (ed.), *Fish Ecology*. (pp. 31-59). New York: Nova Science Publishers.

Rudneva, I.I. and Petzold-Bradley, E. (2001). Environmental and security challenges in the Black Sea region. In: E. Petzold-Bradley, A. Carius and A. Vimce (Eds.), *Environment Conflicts: Implications for Theory and Practice* (pp. 189-202). Netherlands: Kluwer Academic Publishers.

Rudneva, I.I., Kuzminova, N.S. and Omelchenko, S.O. (2012b). Trace elements and nitrozamines concentration in Black Sea elasmobranch species. *Asian J. Biol. Life Sci.*, 1(1), 51-57.

Rudneva, I.I., Kuzminova, N.S., Skuratovskaya, E.N. and Kovyrshina, T.B. (2010b).Comparative Study of Glutathione-S-transferase Activity in Tissues of Some Black Sea Teleosts. *Int. J. Sci. Nature*, 1(1), 1 6.

Solé, M., Baena, M., Arnau, S., Carrasson, M., Maynou, F. and Cartes, J. E. (2010). Muscular cholinesterase activities and lipid peroxidation levels as biomarkers in several Mediterranean marine fish species and their relationship with ecological variables. *Environ. Int.*, 36(2), 202-11.

Sole, M., Rodriguez, S., Papiol, V., Maynou, F. and Cartes, J. E. (2009). Xenobiotic metabolism markers in marine fish with different trophic strategies and their relationship to ecological variables. *Comp. Biochem. Physiol.*, 149 C, 83–89.

Storelli, M. M., Stuffler, R., Storelli, A. and Macotrigiano, G. O. (2005). Accumulation of mercury, cadmium, lead and arsenicin swordfish and blufin tuna from the Mediterranean Sea: A comparative study. *Mar. Pollut. Bull.*, 50, 903-1007.

Svetovidov, A. N. (1964). *Black Sea Fishes*. Leningrad: Nauka. (*in Russian*).

Titov, V. N. and Bochkova, N. A. (1990). Methodological and diagnostic aspects of aminotransferases studies (review). *Laboratory determinations*, 8, 4-12. (*in Russian*).

Turoczy, N. J., Laurenson, L. B., Allinson, G., Nishikawa, M., Lambert, D. F., Smith, C., Cottier, J. P. E., Irvine, B. and Stagnitti, F. (2000). Observations on metal concentrations in three species of shark D*eania calcea*, *Centroscymnus crepidater* and *Centroscymnus owstoni*) from Southeastern Australian Waters. *J. Agric. Food. Chem.*, 48(9), 4357-4364.

Tuteliayn, V. A. (1990). Methods instructions for determination of the Hg in food by the non-flame absorbtion method. *Moscow Inst. of Food RAMS*, 51-78, 97-102.

Van der Oost, R., Beyer, J. and Vermeulen N. P. E. (2003). Fish bioaccumulation and biomarkers in environmental risk assessment: a review. *Environ. Toxicol. Pharmacol.*, 13, 57-149.

Vas, Ph., Gordon, J. D. M., Fielden, P. R. and Overnell, J. (1993). The trace metal ecology of ichthyophauna in the Rockall Trough, north-eastern Atlantic. *Mar. Pollut. Bull.*, 26(11), 607-612.

Velez-Alavez, M., Labrada-Martagon, V., Mendoz-Rodriguez, L., C., Galvan-Magana, F. and Zenteno-Savin, T. (2013). Oxidative stress indicators and trace element concentration s in tissues of mako shark (*Isurus oxyrinchus*). *Comp. Biochem. Physiol. Part A.*, 165 (4), 508-514.

Vysotskaya, R. U. and Nemova, N. N. (2008). *Lysomes and Lisosomal enzymes in fish*. Moscow: Nauka. (in Russian).

Winston, G. W. and Di Giulio, R. T. (1991). Prooxidant and antioxidant mechanisms in aquatic organisms. *Aquat. Toxicol.*, 19, 137-161.

INDEX

C

D

E

F

G

H

I

J

K

L

M

N

O

P

R

S

T

U

V

W

Y

Z